KW-057-988

checkp■int

Endorsed by
University of Cambridge
International Examinations

physics

Peter D Riley

HODDER
EDUCATION
AN HACHETTE UK COMPANY

Titles in this series

Checkpoint Biology Pupil's Book	ISBN 978 0 7195 8067 3
Checkpoint Biology Teacher's Resource Book	ISBN 978 0 7195 8068 0
Checkpoint Chemistry Pupil's Book	ISBN 978 0 7195 8065 9
Checkpoint Chemistry Teacher's Resource Book	ISBN 978 0 7195 8066 6
Checkpoint Physics Pupil's Book	ISBN 978 0 7195 8069 7
Checkpoint Physics Teacher's Resource Book	ISBN 978 0 7195 8070 3

To Kel, Sally and Jenny

LIBRARY & LEARNING CENTRE

CLASS No.	SUPPLIER	PRICE	DATE
530	DAWSON	16.99	12.3.10

ACC. No. B73181

Orders: please contact Bookpoint Ltd, 130 Milton Park, Abingdon, Oxon OX14 4SB. Tel: (44) 01235 827720. Fax: (44) 01235 400454. Lines are open from 9.00–5.00, Monday to Saturday, with a 24-hour message answering service. Visit our websites at www.hoddereducation.co.uk and www.hoddersamplepages.co.uk

© Peter Riley 2005
First published in 2005
by Hodder Education
an Hachette UK Company
338 Euston Road
London NW1 3BH

Impression number	10 9 8 7
Year	2010 2009

All rights reserved. Apart from any use permitted under UK copyright law, no part of this publication may be reproduced or transmitted in any form or by any means, electronic or mechanical, including photocopying and recording, or held within any information storage and retrieval system, without permission in writing from the publisher or under licence from the Copyright Licensing Agency Limited. Further details of such licences (for reprographic reproduction) may be obtained from the Copyright Licensing Agency Limited, Saffron House, 6–10 Kirby Street, London EC1N 8TS.

Cover design by John Townson/Creation
Illustrations by Mike Humphries
Typeset in Garamond 12/14 by Pantek Arts Ltd, Maidstone, Kent
Printed in Italy

A CIP catalogue record for this title is available from the British Library

ISBN 978 0 7195 8069 7

checkp■int

physics

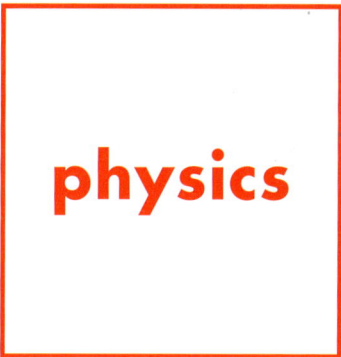

WESTON COLLEGE LEARNING RESOURCE CENTRE
KNIGHTSTONE ROAD
WESTON-SUPER-MARE
BS23 2AL
TELEPHONE: 01934 411493
Fines are payable on overdue items
Return on or before the last date stamped below.

1 0 JAN 2011

1 0 NOV 2014

530
LRC

DAWSON 169912 J 10

B73181

Contents

Preface

To the pupil

Physics is the scientific study of the interactions between matter and energy. These interactions can produce the colours of the rainbow in a shower, or the roar of the wind in a hurricane. At a greater distance, the interactions of matter and energy in the Sun produce light and heat, while inside our eyes light energy is converted to electrical energy, which passes to our brain and allows us to see.

Every event in the Universe, from your next breath to a star exploding, is an interaction of matter and energy, so physics is really a part of all other scientific subjects, rather than a separate one.

Our knowledge of physics has developed from the observations, investigations and ideas of many people over a long period of time. Today physicists – people who study how matter and energy interact – are increasing this knowledge more rapidly than ever before.

In the past, few people other than scientists were informed about the latest discoveries. Today, through newspapers, television and the Internet, everyone can learn about the latest discoveries in a wide range of physics fields, from the exploration of distant planets to the nature of tiny particles inside an atom, and from the development of scanners that allow doctors to see inside the body to satellites that allow us to communicate with other people around the Earth in seconds.

Checkpoint Physics covers the requirements of your examinations in a way that I hope will help you understand how observations, investigations and ideas have led to the scientific facts we use today. The questions are set to help you extract information from what you read and see, and to help you think more deeply about each chapter in the book. Some questions are set so you can discuss your ideas with others and develop a point of view on different scientific issues. This should help you in the future when new scientific issues, as yet unknown, may affect your life.

The scientific activities of thinking up ideas to test and carrying out investigations are enjoyed so much by many people that they take up a career in science. Perhaps *Checkpoint Physics* may help you to take up a career in science too.

To the teacher

Checkpoint Physics has been developed from *Physics Now! 11–14* second edition to specially cover the requirements of the University of Cambridge International Examinations Checkpoint tests and other equivalent junior secondary science courses. It also has the following aims:

- to help pupils become more scientifically literate by encouraging them to examine the information in the text and illustrations in order to answer questions about it in a variety of ways – for example, *For discussion* questions may be used in work on science and citizenship;
- to present science as a human activity by considering the development of scientific ideas from the earliest times to the present day;
- to examine applications of scientific knowledge and issues that arise from them.

This *Pupil's Book* begins with an introduction which briefly reviews the development of science, and in particular physics, throughout the world, then moves on to consider the scientific method and its application. As many consequences of scientific work raise issues, the introduction concludes by addressing an issue with a base in physics, and shows strategies that pupils can use to consider the problems and to try and resolve them. These strategies can be used in the rest of the book in discussion activities where issues with a base in physics arise.

The chapters are arranged in the order of the topics in the Cambridge Checkpoint Physics Scheme of Work with the exception of the topic Measurement, which is Chapter 1. There is an extensive glossary at the back of the book, which includes all the words designated in each topic as essential to the pupil's scientific vocabulary.

This *Pupil's Book* is supported by a *Teacher's Resource Book* that provides answers to all the questions in the pupil's book – those that occur in the body of the chapter and those that occur as end-of-chapter questions.

The resource book also provides end-of-chapter tests, which can be used for extra assessment, and actual questions from past Checkpoint tests. There is a range of practical activities for integration with the work in each chapter to provide opportunities for pupils to develop their skills in scientific investigation.

Acknowledgements

Cover and **p.iii** Tek Image/Science Photo Library; **p.2** *t* Science & Society Picture Library, *b* Topfoto; **p.3** *t* Duncan Shaw/Science Photo Library, *b* Alinari Archives; **p.5** *l* Science Museum/Science & Society Picture Library, *r* Klaus Andrew/Still Pictures; **p.7** Science Photo Library/China Great Wall Industry Corporation; **p.8** Art Archive/Royal Institution/Eileen Tweedy; **p.11** *t* Science Museum/Science & Society Picture Library, *b* © Crown Copyright 1989. Reproduced by permission of the controller of HMSO, photo: National Physical Laboratory; **p.12** *t* Science Museum/Science & Society Picture Library, *bl* Science Photo Library/David Nunuk, *br* Science Photo Library Northwestern University; **p.15** Science Photo Library/Keith Kent; **p.21** *tl* Action Plus/Neil Tingle, *tc* Action Images/Sporting Pictures (UK) Ltd, *tr* Action Plus/Steve Bardens, *bl* Robert Harding Picture Library/Adam Woolfit, *br* Photolibrary.com; **p.22** Action Images/Sporting Pictures (UK) Ltd; **p.24** Science Photo Library/Alex Bartel; **p.27** Andrew Lambert; **p.32** *t* John Townson/Creation, *b* Rex Features/Wil Blanche; **p.34** Hayley Madden/Redferns; **p.35** *t* Getty Images, *b* Still Pictures/Hartmut Schwarzbach; **p.41** Andrew Lambert; **p.42** Andrew Lambert; **p.43** John Townson/Creation; **p.47** Arup Photo Library; **p.50** *t* John Townson/Creation, *b* Robert Harding Picture Library; **p.51** Still Pictures/Mark Edwards; **p.54** John Townson/Creation; **p.55** *t* Andrew Lambert, *b* John Townson/Creation; **p.56** John Townson/Creation; **p.57** Andrew Lambert; **p.58** Andrew Lambert; **p.60** Science Photo Library/Jack Finch; **p.61** *t* Michael Holford, *b* Science Museum/Science & Society Picture Library; **p.64** *t* Science Photo Library/Alex Bartel, *b* Maplin; **p.68** Robert Harding Picture Library/Nigel Francis; **p.69** Science Photo Library/Gordon Garradd; **p.73** *t* Oxford Scientific Films/Matthias Breiter, *b* John Townson/Creation; **p.74** Corbis/© Derek Cattani; **p.82** *t* Science Photo Library/Alexis Rosenfeld, *b* Still Pictures/Xavier Eichaker; **p.85** Still Pictures/H. Verbiesen; **p.86** Alamy/Leslie Garland Picture Library; **p.92** Corbis; **p.93** *t* Getty Images, *b* © Owaki-Kulla/Corbis; **p.95** *l* Alamy/Alt-6, *r* Alamy/Norbert Schaefer; **p.96** *l* Alamy/Holt Studios International Ltd, *r* Alamy/Papilio; **p.97** *t* Zefa, *b* Corbis; **p.99** Alstom; **p.102** *t* Corbis/© Jon Feingers, *b* Science Photo Library/Keith Kent; **p.106** *t* Forest Life Picture Library/Isobel Cameron/© Crown Copyright, *b* Science Photo Library/Dr Jeremy Burgess; **p.107** Action Images/Sporting Pictures (UK) Ltd; **p.108** Science Photo Library/Takeshi Takahara; **p.109** *t* Action Images, *b* Science Photo Library/Detlev Van Ravenswaay; **p.110** *tl* Still Pictures/Jeffrey Rotman, *tr* Oxford Scientific Films/Richard Herrmann, *bl* Robert Harding Picture Library/Martyn F. Chillmaid, *br* Rex Features/Philip Dunn; **p.113** © Crown Copyright, National Meteorological Office; **p.116** Robert Harding Picture Library/Robert Francis; **p.118** Science Photo Library/Detlev Van Ravenswaay; **p.121** John Townson/Creation; **p.123** *t* Last Resort Picture Library, *b* Mary Evans Picture Library; **p.124** Science Museum/Science & Society Picture Library; **p.126** Robert Harding Picture Library/Robert Francis; **p.127** Still Pictures/Kent Wood; **p.128** Science Photo Library/Peter Menzel; **p.131** Andrew Lambert; **p.138** Photodisc; **p.140** Andrew Lambert; **p.144** Action Images; **p.145** *t* Robert Harding Picture Library/Bildagentur Schuster, *b* plainpicture/Alamy; **p.149** Robert Harding Picture Library/Bildagentur Schuster/Herbst; **p.152** Andrew Lambert; **p.153** Science Photo Library/Sheila Terry; **p.154** Robert Harding Picture Library/Adam Woolfit; **p.156** John Townson/Creation; **p.159** *t* Maplin, *b* Andrew Lambert; **p.160** Maplin; **p.163** *tl, tr* Geoscience Features, *b* Shell Photographic Services, Shell International Ltd.; **p.164** Geoscience Features; **p.165** Still Pictures; **p.166** *t* Alamy/Bryan & Cherry Alexander Photography, *b* Still Pictures; **p.167** Still Pictures; **p.168** *l* Mary Evans Picture Library, *r* Science Photo Library; **p.169** Art Archive; **p.170** *t* Martin Bond/Science Photo Library, *b* NASA/Science Photo Library; **p.171** Courtesy of Ocean Power Delivery Ltd (www.oceanpd.com); **p.172** Still Pictures

t = top, *b* = bottom, *l* = left, *r* = right, *c* = centre

Every effort has been made to contact copyright holders but if any have been inadvertently overlooked the Publishers will be pleased to make the necessary arrangements at the earliest opportunity

Introduction

When a young child balances one wooden brick on another, it is making one of its first experiments in physics. It is finding a place for the wooden brick to rest without it being pulled onto the play mat by gravity. Part of the study of physics is concerned with matter and forces like the wooden brick and gravity. The earliest people performed some simple experiments in physics when they chipped away at rock to make sharp, bone-cutting axes.

Children at play.

These simple experiments begin with an observation followed by a test and then another observation and a further test if necessary. For example, the child on the play mat sees that when a brick is placed on top of another one, it often falls off, but when the brick is pushed a little further across the top of the other brick, it stays in place.

A Stone-age person finds a rock with an edge that could be sharpened, chips off a piece of rock and then tests the edge for sharpness by touching it. Throughout human history, people have performed simple experiments in physics without knowing it.

Many of these simple experiments in ancient times led to the making of useful objects such as stone axes, scrapers and knives. About 25 000 years ago people living in a place now called Poland used a mammoth tusk to make a boomerang. When people learned that certain plants could be grown in large numbers to produce food, tools were made to help grow and gather the crop. A simple plough was invented using a stick to scratch a furrow in the soil in which the seeds of the crop plants could be sown. Just over 2500 years ago a plough was

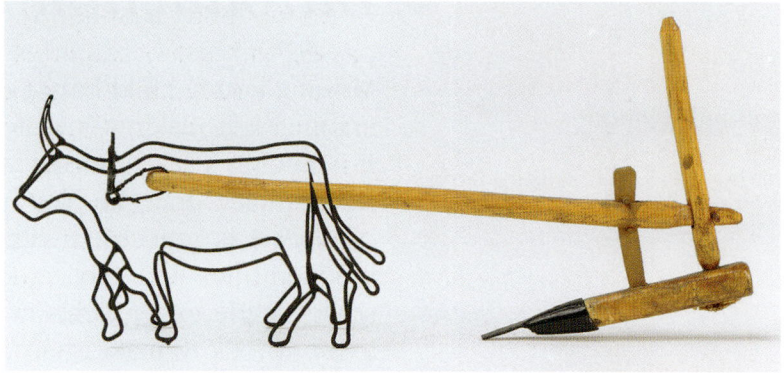

A Chinese plough.

invented that not only cut through the soil but also broke it up and turned it over. This greatly helped crop production and in time became used worldwide.

Science moves on

Once people started farming they could settle down and live in one place. In time, towns and even cities developed. People collected their knowledge of the world and wrote it down. In Greece about 2500 years ago there were places of learning, rather like universities today, which were run by teachers called philosophers. Some Greek philosophers, such as Aristotle, tried to explain observations on the world around them by reasoning and arguing their ideas with others. Aristotle did not test his ideas with investigations. As a consequence some of his conclusions were incorrect. For example, Aristotle believed that if one object was twice as heavy as another it would fall twice as quickly to the ground. Empedocles was another Greek philosopher who came to a wrong conclusion. He concluded that we see things because our eyes send out rays. Archimedes and Hero were philosophers who tested their ideas with investigations. Archimedes studied levers and discovered how they worked. Hero studied how light was reflected from a mirror and found that the angles at which light struck the mirror and reflected off it were the same.

Archimedes stated that if he had a long enough lever he could move the Earth.

About 1200 years ago most of the study of science passed to Muslim countries in Africa and Asia. The Muslim scientists were not content with many explanations of the Greek philosophers and began testing their ideas with scientific investigations. Abu Ali Hasan Ibn al Haitham (965–1040) was also known as Alhazen and performed a great many investigations. He looked at the works of Greeks whose explanations did not fit with observations and put forward more accurate explanations. For example, he considered Empedocles' idea of how we see things to be wrong and explained that we see by the light of the Sun or other luminous objects reflecting the objects around us into our eyes. He also built on the accurate investigations of others. For example, he built up the study of light begun by Hero to explain how lenses work and how a rainbow is created.

A rainbow.

About 400 years ago most of the study of science passed to the countries in Europe. In Italy, Galileo (1564–1642) made a great many investigations from the swinging of pendulums to exploring the surface of the Moon with a telescope. Galileo, like all of the scientists of his day, did not have the measuring and recording instruments we use today such as electronic timers and data loggers. He improvised and developed investigations from his knowledge and the things he had around him. For example, when he observed a chandelier swinging in the cathedral at Pisa, he thought that it took the same time to swing from one side to the other and back again.

A chandelier in Pisa cathedral.

When he got home, he set up a pendulum and as he knew that his resting pulse beat regularly, he used it to check his idea and found it was correct.

Use your pulse as a timer

Try Galileo's investigation by making a long pendulum from a weight on the end of a string and use your pulse to time its swings. Do you find it easy or difficult to do?

Timing rolling objects

Galileo also studied how objects fell. Objects fall quickly and Galileo only had a water clock to measure the time. He reasoned that the pull of gravity, which makes objects fall directly to the ground when they are released, also makes objects roll down slopes. As objects take longer to roll down slopes, especially ones with a very small gradient, Galileo found that he could use his water clock to time the journey made by a ball rolling down a slope.

A water clock can be made from a bowl with a small hole in the bottom. The water that escapes from the hole is collected in another container. Galileo measured time by weighing the amount of water that had collected in the container during his investigation.

Make a water clock

There are many designs for water clocks. Here are some ideas to try.

- You could make one like Galileo's from an old plastic bowl (ask a teacher to make a hole in it for you) and a container to collect the water. You would also need to work out how you were going to weigh the water using a balance.
- You could put a scale in the collecting container and simply measure the water level on the scale when the investigation is complete.
- You may like to put a float with an upright stick and scale in the bowl and measure how far the float sinks below the rim of the bowl.

Try Galileo's experiment

Set up a ramp with a small gradient. Collect two or three balls of the same size but of different masses. Roll each ball down the slope in turn. Time how long each ball takes to make its journey with your water clock. Repeat your experiments to check your results. What do you find?

For discussion
What kinds of research in Physics are being carried out today? Check your ideas by finding out what research is being done at the university nearest to your school. Log on to its website and look for the Physics department web pages.

Issues

Some of the facts that have been discovered through scientific investigation can be used to devise machines to help us survive. For example, the windmill was developed to grind corn and to pump water. It is thought that windmills may have been developed in China and Japan 4000 years ago. It is known for certain that windmills were in use in Iran about 1100 years ago. They had horizontal sails.

About 200 years later, windmills with vertical sails were developed in England and France. From the windmill was developed the wind turbine that is used to generate electrical energy from the moving energy of the wind.

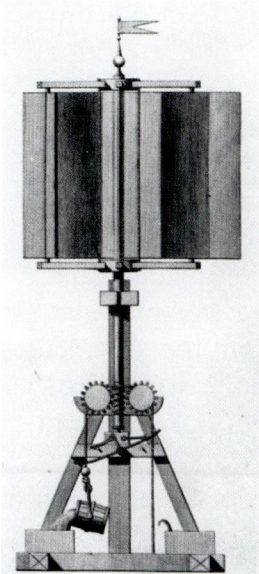

An Iranian horizontal sail windmill.

A wind farm.

Imagine that you lived in a town on the coast. Behind the town were two hills. One hill had no mountains behind it and no one lived there. It was covered in heather and was the home of birds such as the grouse and the skylark. There were a few paths across it for hikers to use. The second hill had mountains behind it and there was a village at its foot. The council (the elected people who ran the town) decided that they could cut down on the use of coal and oil to generate electricity by setting up a wind farm. The problem was where could they site it. It could be on the hill without mountains behind where it would be easily seen from the town, or on the hill with the mountains behind it where the turbines would be less easily seen but their noise might disturb the villagers. Or it could be out at sea where it would not be seen but there would be more expense in bringing the electricity to the town.

When issues like this arise the media (newspapers, radio and television) run reports and articles about them. They often take one view in order to try and influence how people decide to vote on the issue. Imagine that you are a science reporter who is going to write an article giving a balanced view, setting out the advantages and disadvantages of different sites for the wind farm. Research your article by reading more about renewable and non-renewable energy in this book and other sources. Write your article and see if your friends think it is balanced or that you seem to favour one site more than the others.

When looking at other issues in this book you might like to think how a journalist would write an article about them that could give a balanced view of the issue. You may find that there is perhaps not an easy way of presenting a balanced view.

1 Measurement

1 Look at the three lines in Figure 1.2a and write down their letters in order of length, starting with the longest. Repeat the exercise with the lines in Figure 1.2b. When you have finished, check your answers by measuring the lines. What does this tell you about your senses and the need to make measurements?

a)

A ————————————

B ——————————

C ——————————

b)

A ⟩————————⟨

B ⟨————————⟩

C |————————|

Figure 1.2

2 A girl puts her left hand in a bowl of cold water and her right hand in a bowl of hot water. After a minute she puts both hands in a bowl of warm water. How do you think the left hand and the right hand will feel in the bowl of warm water?

Launching a space rocket

Figure 1.1 A multi-stage rocket leaving the launch pad.

Three, two, one, zero, lift off! Light from the rocket engines can be seen immediately by the distant spectators as the rocket begins to rise from the launch pad. When the roar of the rocket engines reaches the spectators it nearly deafens them. The rocket's speed increases every second as it rises into the sky.

The rocket is divided into parts, called stages. Each stage has fuel tanks and rocket engines. When the fuel is used up in one stage that stage will separate from the rocket and fall back towards Earth. As the stage rushes back through the atmosphere it will become so hot that it will burn up. When the last stage has separated only a small spacecraft will remain in orbit around the Earth or set off across the Solar System.

Some of the things that are described in the first two paragraphs are called phenomena. Each phenomenon, such as light and sound, can be investigated. The science of investigating phenomena such as light, sound and forces is called physics.

Scientists begin an investigation by asking a question. They think of an explanation to answer the question and then test their explanation with an experiment.

A major part of an investigation is the making of observations. We use our senses to make observations – but our senses can be unreliable.

For discussion

After reading about the rocket launch, a person asked, 'Why was light from the rocket seen before the sound of the rocket was heard? Why did the stages fall back to Earth when they separated from the rocket? Why did the stages burn up in the atmosphere?'

What explanations can you give to answer these questions?

Length, mass and time

Rather than relying on senses, more accurate observations of phenomena are made by taking measurements. Three things that are measured in many investigations are length, mass and time. Any measurement is made in units, for example a common unit of length is the centimetre. There is an international system of units that is used by scientists throughout the world. This is known as the Système International d'Unités. The units in this system are known as SI units.

Measuring length

The standard SI unit of length is the metre. Its symbol is m, and, as with all symbols of SI units there is not a full stop placed after it. The metre is divided into smaller units for measuring small lengths or distances, and large

Explaining phenomena

In the period from 600 to 300BC, the Ancient Greek philosophers explained many phenomena but did not test their explanations with experiments. They made observations, formed opinions and argued about what they saw. The opinions of the Ancient Greeks were taught as facts for 2000 years.

Francis Bacon (1561–1626), an English philosopher, disagreed with the way the Ancient Greeks had explained phenomena. He set out the process of investigation that we use today. In this process, he said, the investigator should choose the facts to be investigated, form a hypothesis (an idea that links the facts together and can be tested), perform the test and evaluate the result; investigations should be repeated and from their results general theories and laws could be set up to explain the phenomena that are being studied. This change in how phenomena should be investigated became popular and eventually led to the setting up, in 1660, of the Royal Society in London where scientists could report on their investigations and discuss their work. Later, in 1799, the Royal Institution was set up where scientists demonstrated their experiments to the public.

1 Which of the following statements contain
 a) facts that can easily be investigated,
 b) an opinion that cannot easily be investigated?
  i) I am stronger than you.
  ii) I am the best person in the world.
  iii) You can only see the Moon at night.
  iv) Chips are nicer than crisps.
Explain your answer in each case.

Figure A   Sir James Dewar (see page 135) lecturing on liquid nitrogen at the Royal Institution.

numbers of metres are made into bigger units to measure long lengths or distances. Table 1.1 shows some of these other SI units.

Table 1.1 Units of length.

Unit	Symbol	Number of metres
kilometre	km	1000 m
metre	m	1 m
centimetre	cm	0.01 m
millimetre	mm	0.001 m
micrometre	µm	0.000 001 m
nanometre	nm	0.000 000 .001 m

3 What is the mass of the object on each of the balances shown in Figure 1.3? Which was the easiest to read?

Measuring mass

The standard SI unit of mass is the kilogram, whose symbol is kg. The other SI units of mass used in investigations are shown in Table 1.2.

Table 1.2 Units of mass.

Unit	Symbol	Number of kilograms
Megatonne	Mt	1 000 000 000 kg
tonne	t	1000 kg
kilogram	kg	1 kg
gram	g	0.001 kg
milligram	mg	0.000 001 kg

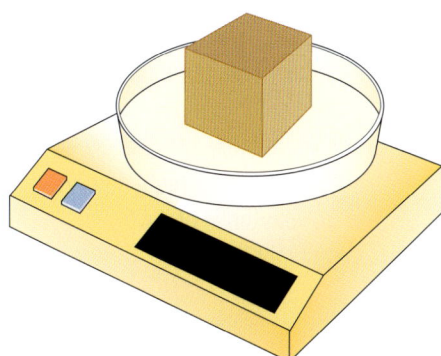

top-pan balance

weighing scales

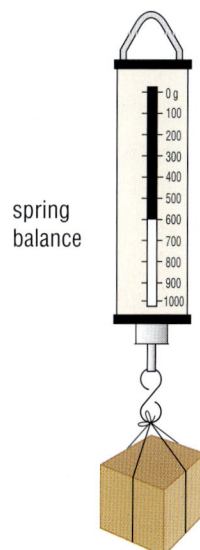

spring balance

100 g

Figure 1.3 Measuring mass.

Measuring time

The standard SI unit of time is the second and its symbol is s. Other units of time used in investigations are shown in Table 1.3.

Table 1.3 Units of time.

Unit	Symbol	Number of seconds (minutes or hours)
day	d	86 400 s (1440 minutes, 24 hours)
hour	h	3600 s (60 minutes)
minute	min	60 s
second	s	1 s
millisecond	ms	0.001 s

4 If you saw someone commit a crime, how might you describe to a detective the appearance of the criminal and what happened? Do mass, length and time feature in your answer? If they do, say where they occur.

Finding a standard

If measurements are to be useful to a large number of people, the same units must be used by everyone. At first many people used parts of their own bodies as units of measurement. Just as horses are measured in 'hands' today, the Ancient Egyptians around 2000 to 3000BC measured small distances by parts of the hand and arm. One width of the finger was a digit, four digits made a palm, five digits made a hand and 28 digits were a cubit, which was the distance from the fingertip to the elbow. However, as people vary in size the units also varied from one person to the next and led to confusion. In a city in Mesopotamia about 4000 years ago the people removed this confusion by basing their unit of measurement on the foot of a statue of the city's governor.

People measured longer distances, such as those in journeys, in units of time. The journeys were measured in hours, days or even 'moons'.

The first measure of time was the period of daylight and the period of dark. These two periods of time were then each divided up into 12 sections called temporal hours. The first hour of daylight began with the dawn and the first hour of the night began with the sunset. As the time of dawn and sunset varied through the year, the length of the hours of day and night varied too and depended on the time of year. Later the time taken for the Earth to complete one rotation, from midday to the following midday, was divided up into 24 one-hour periods.

The passing of the daylight hours was first measured with a shadow stick, invented in China about 4500 years ago. It was set upright in the ground and the change in position of the shadow was used to tell the time of day. About 1000 years later the Egyptians invented the sundial, which had a tilted bar called a gnomon that cast a shadow across a surface on which the hours were marked. Water clocks, sand clocks and candle clocks were used to measure time during the night.

1 a) Are all your digits on one hand the same width?
  b) Does this affect your measurement of a short length under four digits? Explain your answer.
2 Is your cubit 28 digits?
3 Why was confusion over measuring removed by using the foot of a statue as a standard length?
4 What time period is a 'moon'?
5 In the first method of time-keeping, how would the length of the 12 temporal hours of
  a) a summer's day compare with those of a winter's day,
  b) a summer's night compare with those of a winter's night?

The invention of mechanical clocks in the 1650s led to a closer synchronising of measured time with the turning of the Earth.

Jean Picard (1620–1682) was a French astronomer. He used his telescope and a set of measuring instruments to measure angles between the Earth and the stars. From his measurements he was able to calculate the circumference of the Earth.

Figure A Some sand clocks consisted of sets of sand glasses. Each one measured a different period of time.

In 1789 the French Revolution began. It brought great change to the way France was run. In 1790 a group of French scientists met to decide on standards that could be used throughout the country. They thought that the standard unit of length should be a fraction of the Earth's circumference. They decided that it should be one ten-millionth of the distance between the north pole and the equator along a north–south line. This unit of length was called the metre. They also decided that the standard unit of mass should be the gram and be based on the mass of a certain volume of water at a certain temperature. The system that they set up is called the metric system.

In time other countries adopted the metric system for all their measurements. In 1875 an International Bureau of Weights and Measures was set up in Paris. The people working there reviewed the standards and decided to replace them. Measurement of the metre based on the Earth's circumference was difficult to check so the standard for the metre became a distance between two marks on a bar of platinum–iridium alloy which was much easier to examine. The standard for mass was also made easier to check. It became a thousand times bigger – the kilogram, and is based on the mass of a platinum–iridium alloy block called the international prototype kilogram.

6 How was the Earth used to provide the standards of time and length?
7 Why is the Earth no longer used to provide these standards?
8 Two of today's standards can be measured by scientists anywhere in the world provided they have the equipment. Which standards are they?
9 How does having standard units of measurement help scientific investigation?

Figure B The international prototype kilogram. *(continued)*

11

Later the original standards of time and length were found to be unreliable. For example, further studies on the turning of the Earth showed that over the first half of the 20th century its speed of daily rotation had slowed by 2 seconds. This led to the search for another standard for measuring time and in 1967 it was decided that a second should be that period of time in which the microwaves produced by hot caesium atoms vibrate 9 192 631 770 times.

Today the standard length of a metre is the distance covered by light in a vacuum in 1/299 792 458th of a second.

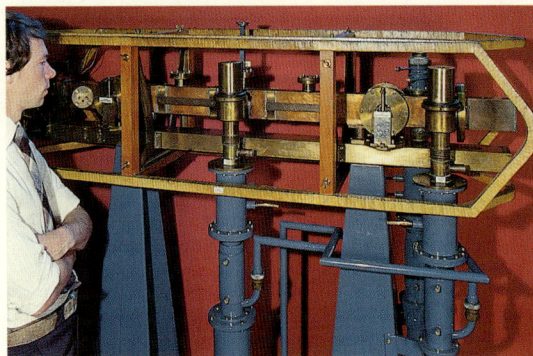

Figure C   An 'atomic clock' in which the microwaves produced by caesium atoms are measured.

Very large and very small numbers

When very large or very small numbers are written down they are arranged in groups of three from the decimal point, without a comma between the groups.

For example, the Moon is 384 000 000 metres from the Earth and the 'width' of an atom is about 0.000 000 001 metres.

Figure 1.4   The Moon is 384 000 000 metres from the Earth.

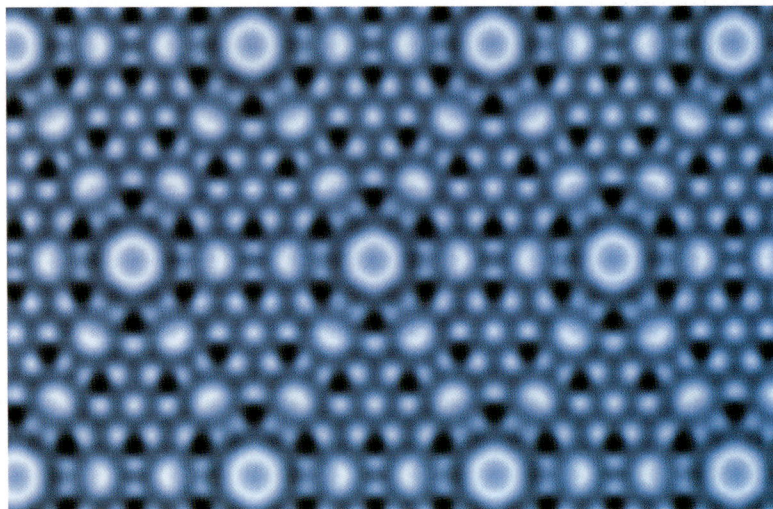

Figure 1.5   The arrangement of individual atoms on the surface of a silicon crystal can be seen with an electron microscope.

5 It is one thousand four hundred and twenty-seven thousand million kilometres from the Sun to Saturn. Write this down
  a) as an ordinary whole number,
  b) in standard form (see page 13).

These very large and very small numbers can be more quickly written down and understood by using powers of ten. This saves writing down or interpreting large numbers of noughts. The following examples show how powers of ten can be used.

6 A bacterium feeding on a decaying body in the soil has a length of one-millionth of a metre. Write this down
   a) as a decimal fraction,
   b) in standard form.

7 Rewrite the following numbers in standard form:
   a) 780×10^3,
   b) 49×10^6,
   c) 247×10^{-6},
   d) 8032×10^{-3},
   e) 0.548×10^4.

8 Estimate these quantities, then check your answers by measurement:
   a) the length of your index finger and the length of your thumb,
   b) the height of your chair,
   c) the distance between you and a door,
   d) the mass of
      i)   this book,
      ii) your school bag and its contents,
   e) the time it takes you to
      i)   count the first 50 words under *Accuracy of measurements*,
      ii) say those 50 words,
      iii) write down those 50 words.

9 a) How could you estimate the thickness of a page of this book?
   b) Write down your estimate of the thickness of a page of this book and compare it with the estimates made by others. Are all the estimates the same? Explain what you discover.

- 20 is 2×10 so it can be written as 2×10^1
- 200 is $2 \times 10 \times 10$ so it can be written 2×10^2

The distance between the Earth and the Moon is $384 \times 10 \times 10 \times 10 \times 10 \times 10 \times 10$ metres, which can be written as 384×10^6 metres.

In order to make the system of recording large numbers consistent in science, only one figure is written in front of the decimal point. This way of writing the figure is called the standard form. The standard form of 384×10^6 is 3.84×10^8.

Very small numbers have a number of noughts after the decimal point. They can also be written down using powers of ten. Each time the decimal point is moved one nought or number to the right the power of ten changes by 10^{-1}. For example:

- 0.1 is 1.0×10^{-1}
- 0.01 is 1.0×10^{-2}

The width of an atom is about $0.000\,000\,001$ metre, which is 1.0×10^{-9} metre.

Estimating quantities

At the beginning of an investigation it may be useful to estimate the quantities that are going to be used or the time that is going to be taken for certain observations. At this stage of the investigation accuracy is not essential – that comes later.

Accuracy of measurements

Your accuracy when making a measurement depends on the measuring instrument – how well it has been made, calibrated (compared with the standard) and how well the scale on the instrument has been constructed. A stopclock which only measures time by seconds cannot be used to time events to a tenth of a second. Your accuracy also depends on how well you use the measuring instrument. Care in setting up the device is needed. This includes placing a ruler accurately (both ends are important when measuring length), resetting a stopclock before repeating a timing and making sure that a balance is set at zero before a mass is put on it. If a balance is used with a scale that is read by looking at the position of a pointer, your eye should be placed directly in front of the pointer.

10 In Figure 1.6, how would looking at the pointer from positions A and B affect the accuracy of the measurement?

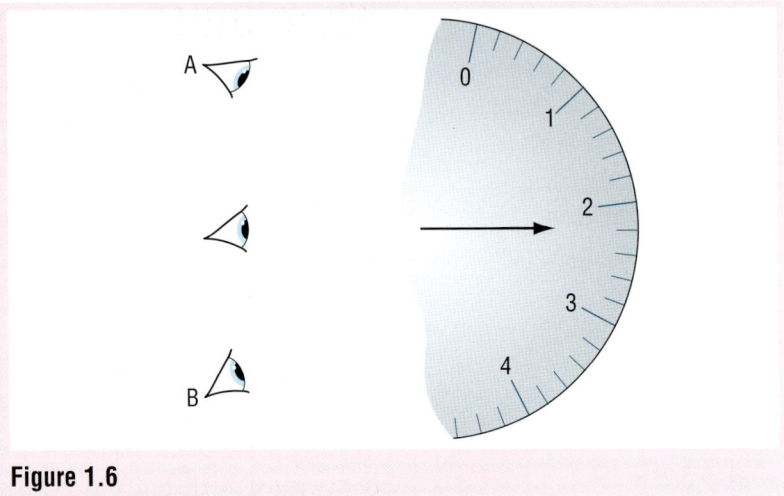

Figure 1.6

The accuracy you claim for a measurement is shown in the way that you write it down, as these examples show:

- A length written down as 10 cm means that the length may be 9.5 cm or 10.5 cm. The difference between the two extremes is 1.0 cm.
- A length written down as 10.0 cm means that the length may be 9.95 cm or 10.05 cm. The difference between the two extremes is 0.1 cm.
- A length written down as 10.00 cm means that the length may be 9.995 cm or 10.005 cm. The difference between the two extremes is 0.01 cm.

11 A length is shown as 10.55 cm.
  a) What is the shortest and the longest length it could be?
  b) What is the difference between these two possible lengths?
  c) How does the accuracy of this reading compare with the 10.5 cm claimed in Figure 1.7?

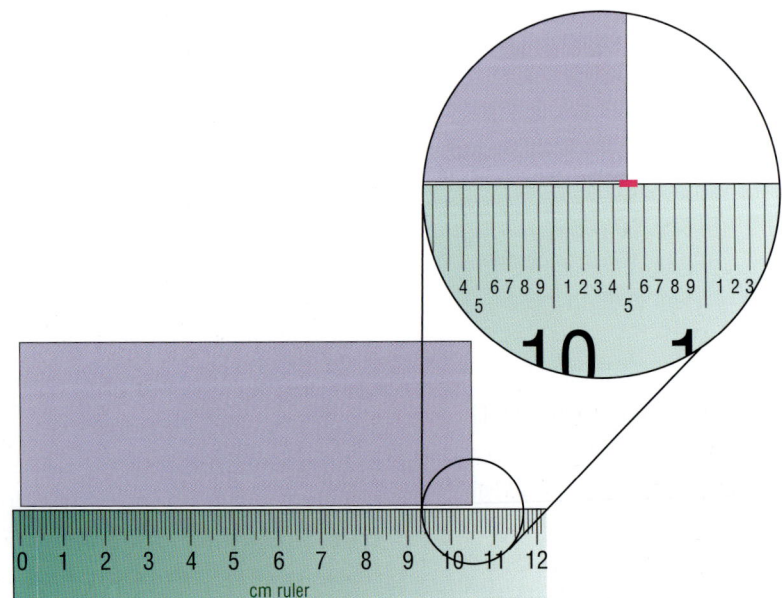

Figure 1.7 If you measure a length and claim it to be 10.5 cm, you are claiming that the ruler and your eyesight are good enough to be sure the edge of the object is in the range 10.45 cm to 10.55 cm.

More units

The units for measuring area and volume are built up from the units of length.

Area

An area is a measure of a surface. The area of the surface of a rectangular shaped object can be found by multiplying the length of the side by the width. The length and width must be measured in the same units – that is both are measured in metres or centimetres or millimetres. The quantities used to find an area are both length quantities so the area of a surface is a length multiplied by a length, or length squared. The standard SI unit of area is m^2 (square metre) but you may also use cm^2 or mm^2 depending on the length unit which is most convenient. An area of land may be measured in km^2.

Figure 1.8   Measuring an area at an archaeological site.

12 What is your area in contact with the surface of the Earth when you stand up? Estimate the area then find out by measuring the area of the soles of your shoes. Which unit is it easiest to use in doing this?

Many areas are irregularly shaped. They may be measured by drawing the outline of the area on squared paper and counting the number of squares and fractions of squares inside the outline.

Volume

The volume of an object is the amount of space that it occupies.

Measuring the volume of a regular shape

The volume of a solid with a regular shape, such as a cube, is found by multiplying together the length, height and width of the cube. All three quantities are lengths so the volume is a length × a length × a length, or length cubed. The volume of a cube with sides of one metre is $1 \, m^3$, or one cubic metre.

This method of multiplying length, width and height can also be used to calculate the volume of rectangular blocks.

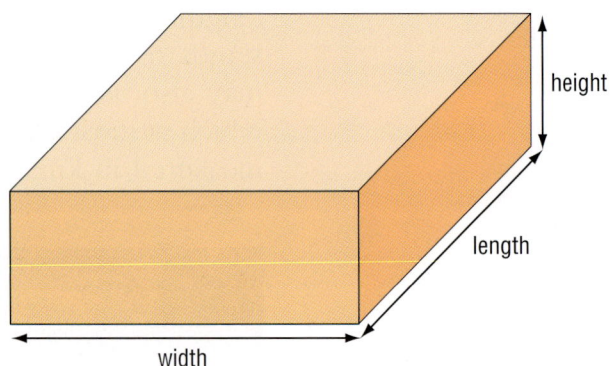

Figure 1.9 The quantities for calculating volume.

Measuring the volume of a liquid

The volume of a liquid can be found by pouring it into a measuring cylinder and reading the volume of the liquid from the scale. (See also *Checkpoint Chemistry*.) Care must be taken to read the scale level with the horizontal part of the liquid surface and not the curved part (meniscus) which forms where the liquid meets the side of the cylinder. Most liquids curve downwards but mercury curves upwards.

Measuring the volume of an irregularly shaped solid

An irregularly shaped solid, such as a pebble, does not have sides that can be easily measured. Its volume can easily be found by using a liquid. Water is poured into a measuring cylinder until the cylinder is about half full. The volume of the water is measured, then the pebble is lowered gently into it. When the pebble is completely immersed the volume of the water is read again. The volume of the pebble is found by subtracting the first reading from the second.

13 What is the volume of a solid block measuring 3 metres long by 2 metres wide and 4 metres high?

14 What is the volume of your bedroom? Estimate the volume first then make measurements and calculations. How good was your estimate? Which units did you use?

15 What is the volume of liquid in cylinders **a)** and **b)** in Figure 1.10?

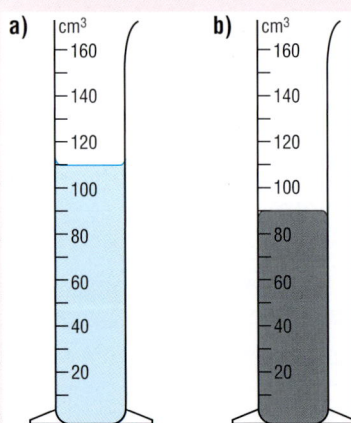

Figure 1.10

16 Find the volume of a pebble from these readings:
Original volume of water in cylinder = 50 cm^3.
Combined volume of water and pebble = 84 cm^3.

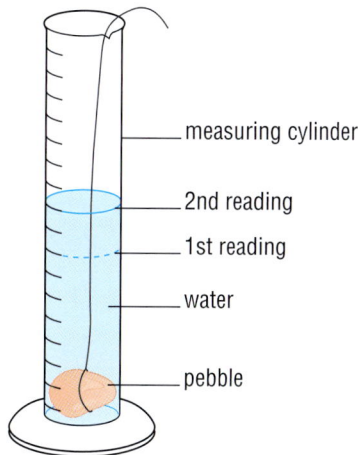

Figure 1.11 Measuring the volume of a pebble.

17 How much hotter is
  a) 45 °C than 30 °C,
  b) 20 °C than −15 °C?

18 Why are two fixed points needed for a temperature scale and not just one?

Heat and temperature

The hotness or coldness of a substance is measured by taking its temperature. The temperature of a substance is measured on a scale which has two fixed points. The most widely used temperature scale is the Celsius scale. Its two fixed points are 0 °C (the melting point of ice or freezing point of water) and 100 °C (the boiling point of water). In between the two fixed points the scale is divided into 100 units or degrees. The scale may be extended below 0 °C and above 100 °C; laboratory thermometers usually have a scale reading from −10 °C to 110 °C.

The thermometer compares the temperature of the substance in which the bulb is immersed with the freezing point and boiling point of water. It compares the hotness or coldness of a substance. It does not measure the total internal energy (see page 30) of the substance.

The lowest possible temperature, known as absolute zero, is −273 °C. Temperatures can go as high as millions of degrees Celsius.

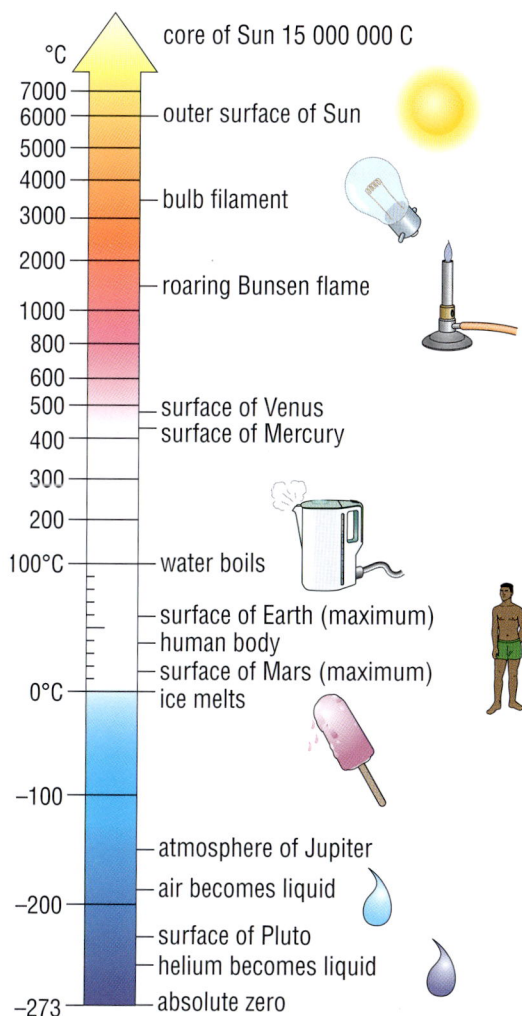

°C
core of Sun 15 000 000 C
7000
6000 — outer surface of Sun
5000
4000
3000 — bulb filament
2000
— roaring Bunsen flame
1000
800
600
500 — surface of Venus
400 — surface of Mercury
300
200
100°C — water boils
— surface of Earth (maximum)
— human body
— surface of Mars (maximum)
0°C — ice melts
−100
— atmosphere of Jupiter
— air becomes liquid
−200
— surface of Pluto
— helium becomes liquid
−273 — absolute zero

Figure 1.12 The Celsius scale of temperature.

19 Which thermometer, one containing mercury or one containing alcohol, could be used in a polar region where the temperature reaches below −40 °C? Explain your answer.

20 Which type of thermometer could be used to measure the boiling point of water? Explain your answer.

Liquids in thermometers

Two liquids that are commonly used in thermometers are mercury and alcohol. Mercury has a freezing point of −39 °C and a boiling point of 360 °C. Alcohol has a freezing point of −112 °C and a boiling point of 78 °C.

◆ SUMMARY ◆

- ◆ Better observations are made by taking measurements, and an international system of units has been developed (*see page 8*).
- ◆ Powers of ten are used for dealing with very large and very small numbers (*see page 12*).
- ◆ It is useful to estimate quantities at the beginning of an investigation (*see page 13*).
- ◆ There are techniques for accurate measuring (*see page 13*).
- ◆ Units of area and volume are built up from units of length (*see pages 15–16*).
- ◆ Temperature is a measure of the hotness or coldness of a substance (*see page 17*).
- ◆ The expansion and contraction of liquids in thermometers are used to measure temperature (*see this page*).

End of chapter questions

1   A model rocket was launched and measurements of its height and horizontal distance from the launch pad were taken as shown in the table.

   a) Plot the flight path of the model rocket from these measurements.

   b) What was the vertical distance from the launch pad when the rocket stopped rising?

   c) What was the horizontal distance from the launch pad when the rocket stopped rising?

   d) What was the horizontal distance from the launch pad when the rocket hit the ground?

   e) If the rocket had been 10 metres above the ground when its horizontal distance from the launch passed was 2 metres, where do you think the rocket would have landed?

Horizontal distance from launch pad/m	Vertical distance from launch pad/m
0	0
1	4
2	8
3	11
4	13
5	14
6	15
7	15.5
8	15
9	13
10	10
11	0

2   A volume of liquid was heated and cooled. During this time the temperature of the liquid was measured nine times.

   a) Plot a graph of the data in the table.

   b) What do you think the temperature was at
   i) 5 minutes, ii) 7 minutes?

   c) Did the liquid heat up and cool down at the same rate? Explain your answer.

   d) What would you estimate the temperature to be after 13 minutes?

Time/min	Temperature/°C
0	20
1	40
2	60
3	80
4	90
6	96
8	80
10	60
12	40

2 *Forces and their effects*

Forces

1 Describe the pushing and pulling forces shown in Figure 2.1a–h.

You cannot see a force but you can see what it does. You can also feel the effect of a force on your body. A force is a push or a pull.

Figure 2.1 Forces act in many ways.

What forces do

- A force can make an object move. For example, if you throw a basketball your muscles exert a pushing force on the ball and it moves through the air when you let it go.
- A force can make a moving object stop. For example, a goalkeeper moves into the path of a moving ball to exert a pushing force on the ball to stop it.
- A force can change the speed of a moving object. For example, a hockey player uses a hockey stick to push a slow-moving ball to send it shooting past a defender.
- A force can change the direction of a moving object. For example, a batsman can change the direction of a cricket ball moving towards the wicket by deflecting it so that it moves away from the wicket towards the boundary.
- A force can change the shape of an object. For example, when a racket strikes a tennis ball, part of the ball is flattened before the ball leaves the racket.

Figure 2.2 A force makes the ball move.

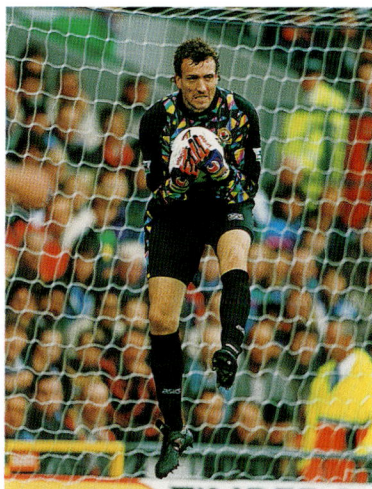

Figure 2.3 A force makes the ball stop.

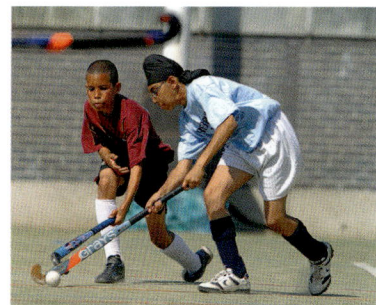

Figure 2.4 A force increases the speed of the ball.

For discussion

Watch a short video recording of part of a football or tennis match and identify the different effects of forces acting in the game.

Figure 2.5 A force changes the direction of the ball.

Figure 2.6 A force changes the shape of the ball.

21

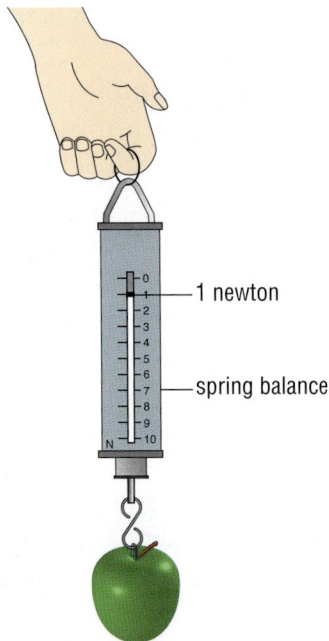

1 newton

spring balance

Figure 2.7 The apple pulls on the spring balance with a force equal to its weight.

How to measure a force

A force can be measured with a Newton spring balance (see page 26). The SI unit for measuring force is the newton (symbol N). This force is quite small and is equal to the gravitational force on (the weight of) an average-sized apple, or the pulling force needed to peel a banana!

Different types of forces

There are two main types of forces: contact forces and non-contact forces. A contact force occurs when the object or material exerting the force touches the object or material on which the force acts. A non-contact force occurs when the objects or materials do not touch each other.

Contact forces

All the situations described so far are examples of contact forces in action. Some more examples follow.

Impact force

When a moving object collides with a stationary object an impact force is exerted by one object on the other. The size of the force may be large such as when a hammer hits a nail or it may be very tiny such as when a moving molecule of gas in the air strikes the skin.

Figure 2.8 The car behind exerts an impact force on the car in front.

2 a) i) What do you feel if you hook the two ends of an elastic band over your index fingers and slowly move your hands apart?

ii) What happens to the elastic band when you bring your hands together again?

iii) What happens if one end of a stretched elastic band is released?

b) Describe how the strain force changes in each part of question a).

Strain force

When some materials are squashed, stretched, twisted or bent they exert a force which acts in the opposite direction to the force acting on them. These materials are called elastic materials and the force they exert when they are deformed is called a strain force. When the force applied to the material is removed the strain force exerted by the material restores the deformed material to its original shape. For example, the strain force in the squashed tennis ball in Figure 2.6 returns the ball to its original shape when the ball has left the racket.

Tension is a strain force that is exerted by a stretched spring, rope or string. At each end the tension force acts in the opposite direction to the pulling force. (See *How to measure a force*, page 22, and *Balanced forces*, page 112.)

A force is shown in a diagram as an arrow pointing in the direction of the push or the pull.

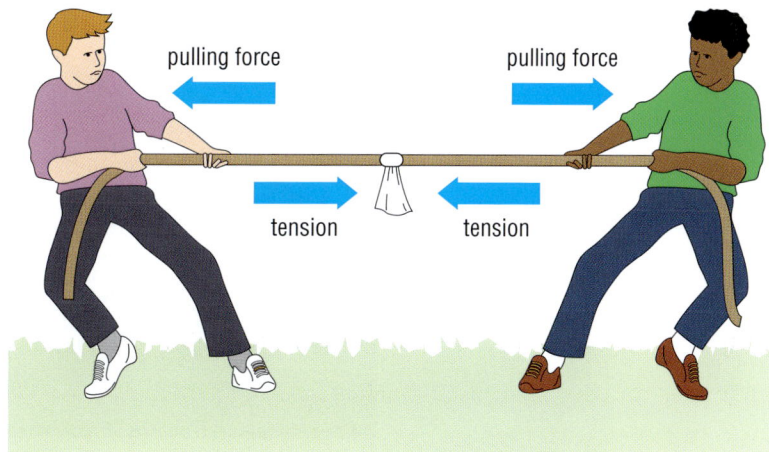

Figure 2.9 Tension is a strain force which acts against the force applied.

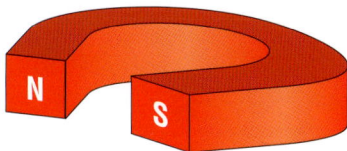

Non-contact forces

These include magnetic forces, electrostatic forces and gravitational forces. They all exert their force without having to touch the object.

Magnetic force

A magnet has a north-seeking pole and a south-seeking pole. These are usually known as the north pole and the south pole (see page 55). If you pick up two magnets and bring either their north poles or their south poles together you will feel a force pushing your hands apart as the two similar poles repel each other. You will feel your hands being pushed away even though the magnets

N = north pole
S = south pole

Figure 2.10   Bar, horseshoe and ring magnets.

are not touching. The strength of the push increases as you bring the two similar poles closer together.

If you bring the north pole of one magnet towards the south pole of another magnet you will feel your hands being pulled together as the different poles attract each other. The strength of this pull increases as the poles get closer together.

A magnet can also exert a non-contact force on objects made of iron, steel, cobalt or nickel. Either pole of the magnet exerts a pulling force on these magnetic materials. The strength of the force increases as the magnet and the magnetic material are brought closer together.

3 If you had a magnet with its north and south poles marked on it and a magnet without its poles marked, how could you identify the poles of the unmarked magnet? Explain your answer.

Figure 2.11 This 'Maglev' train is supported above its track by strong magnetic forces. It travels quietly on a 'cushion' of air which eliminates friction between the train and the tracks.

Electrostatic force

As we shall see in Chapter 11 (page 121), if certain electrical insulator materials are rubbed an electrostatic charge develops on them. There are two kinds of charge: positive charge and negative charge. The forces between the charges can be investigated by suspending a plastic rod so that it can swing freely (see Figure 2.12), giving the rod an electrostatic charge and then bringing rods with different charges close to it. If the suspended rod has a positive charge it will move away from a plastic rod which also has a positive charge, as the similar charges repel each other. If a rod with a negative

4 Compare magnetic and electrostatic forces. In what ways are they
a) similar,
b) different?

charge is brought near the positively charged suspended rod, the rod swings towards it as the opposite charges attract each other. The strength of the force between electrostatic charges increases as the rods are brought closer together.

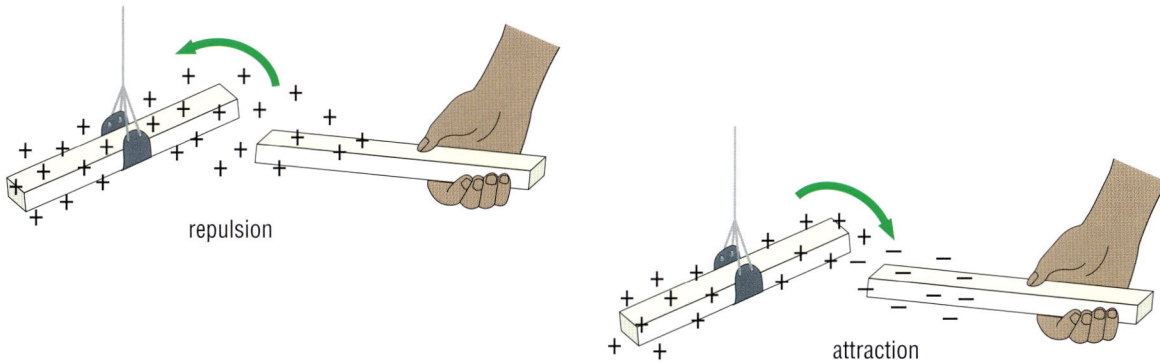

Figure 2.12   Investigating electrostatic force with charged plastic rods.

Gravitational force

There is a force between any two masses in the Universe. The masses may be small such as those of an ant and a pebble or they may be very large such as those of the Sun and the Earth. The force that exists between any two masses because of their mass is called the gravitational force. The force acting between small masses is too weak to have any noticeable effect on them but the gravitational force between two large masses such as the Sun and the Earth is large enough to be very important. It is the gravitational force between the Sun and all the planets in the Solar System that holds the planets in their orbits (see page 117). The gravitational force between an object on the Earth and the Earth itself pulls the object down towards the centre of the Earth and is called the weight of the object.

How springs stretch

Robert Hooke (1635–1703) investigated the way in which springs stretched when masses were attached to them. He first hung up a spring and measured its length without any mass attached to it. He then hung a mass on the bottom and measured the new length of the spring. He calculated the extension of the spring by subtracting the original length of the spring from the new length of the spring with the mass attached. Hooke

5 A spring is 6 cm long when it is unstretched but is stretched to 9 cm when a mass is hung from it. What is the extension of the spring?

6 An unstretched spring is 6 cm long but becomes 7 cm long when a 100 g mass is hung from it. The spring becomes 8 cm long when a 200 g mass is hung from it.

  a) What is the extension for each mass?

  b) What extension do you predict when masses of i) 300 g and ii) 350 g are hung from it in turn? Can you be sure that the extension values you predict will in fact occur? (Hint: think about the elastic limit.)

repeated the experiment with different sizes of masses. Each time he found the total extension by subtracting the original length from the new length. He found that as the size of the mass increased the size of the extension increased in proportion: the extension of the spring was proportional to the mass attached to it.

Each time Hooke removed the mass the spring returned to its original length. However, he eventually placed a mass on the spring that stretched the spring so much that it remained slightly stretched when the mass was removed. The spring had gone beyond a point called the elastic limit and was permanently deformed. When a larger mass was then added to the spring it no longer extended in proportion to the mass. The spring beyond its elastic limit was in a state known as plastic deformation (see Figure 2.13).

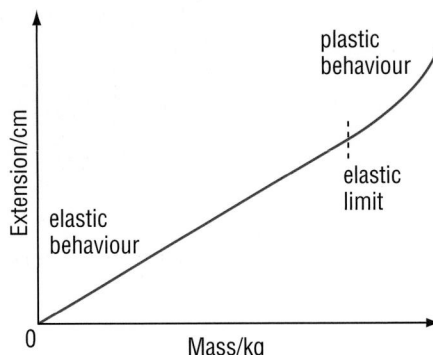

Figure 2.13 How the extension of a spring varies with the mass attached to it.

The Newton spring balance

The discovery made by Robert Hooke has led to the development of a force measurer using a spring which is not stretched beyond its elastic limit. This instrument is called a spring balance. The extension of the spring, and hence the reading on a scale, is proportional to the weight of the mass hung from it, or the force with which it is pulled. The scale of the balance is calibrated in newtons so it is sometimes called a Newton spring balance or a newtonmeter.

There is a range of spring balances which measure forces of different sizes. For example, a spring balance may measure forces with values in the range 0–10 N, 0–100 N or 0–200 N.

There is a device called a stop on most spring balances. It prevents the spring from stretching beyond its elastic limit.

7 How do you think the spring in a spring balance with a scale of 0–10 N compares with a spring in a spring balance that measures forces up to 500 N?

8 A spring balance without a stop would not give correct readings for the weights of the masses hung from it if large masses were used. Explain the reason for this.

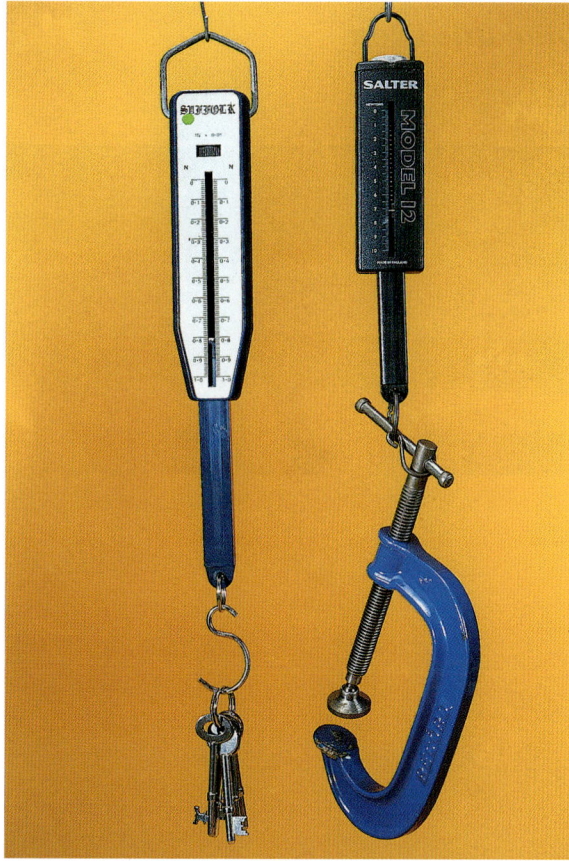

Figure 2.14 Spring balances with different scales.

◆ SUMMARY ◆

◆ A force is a push or a pull (*see page 20*).

◆ A force can make an object move, stop, change speed or change direction or can change an object's shape (*see page 21*).

◆ A force can be measured by a Newton spring balance (*see page 22*).

◆ Two contact forces are impact forces and strain forces (*see pages 22–23*).

◆ Three non-contact forces are magnetic forces, electrostatic forces and gravitational forces (*see pages 23–25*).

◆ Springs stretch when forces are applied to them (*see page 25*).

End of chapter question

1 Identify the forces acting in this scene.

Figure 2.15

3 Fuels and energy

What is energy?

Energy exists in many forms. In the following pages the main forms of energy are described.

Figure 3.1 Energy exists in many forms.

For discussion

Think about conversations where you might use the word energy. What do you mean when you use the word?

In everyday language we use the word energy in many different ways but in science it only has one meaning. The scientific way of thinking about energy is to say that it is the property of something that makes it able to exert a force and do work. To understand this it is helpful to think about the ways that energy is stored and what happens when it is changed from one form to another.

Forms of energy

Heat energy

This is also called thermal energy or internal energy. All substances are made up of particles. They possess a certain amount of energy which allows them to move. When a substance is heated this movement increases. For example, the particles in a solid are moving backwards and forwards about a fixed position. The particles in a liquid move more quickly and can move past each other. The particles in a gas can move freely in all directions at high speeds. When a substance is heated, the particles receive more energy and move faster.

Radiation energy

There is a form of energy that can travel through space at the speed of light. This kind of energy travels in waves that have some properties of electricity and some properties of magnetism. They are called electromagnetic waves. There is a huge range of possible wave sizes, or wavelengths.

Electromagnetic waves are split into seven groups according to wavelength, as Figure 3.3 shows. The different groups have different properties and different uses. The two most familiar groups are light and radio waves.

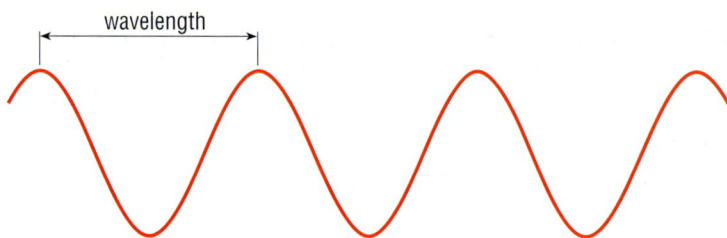

Figure 3.2 A wave showing wavelength.

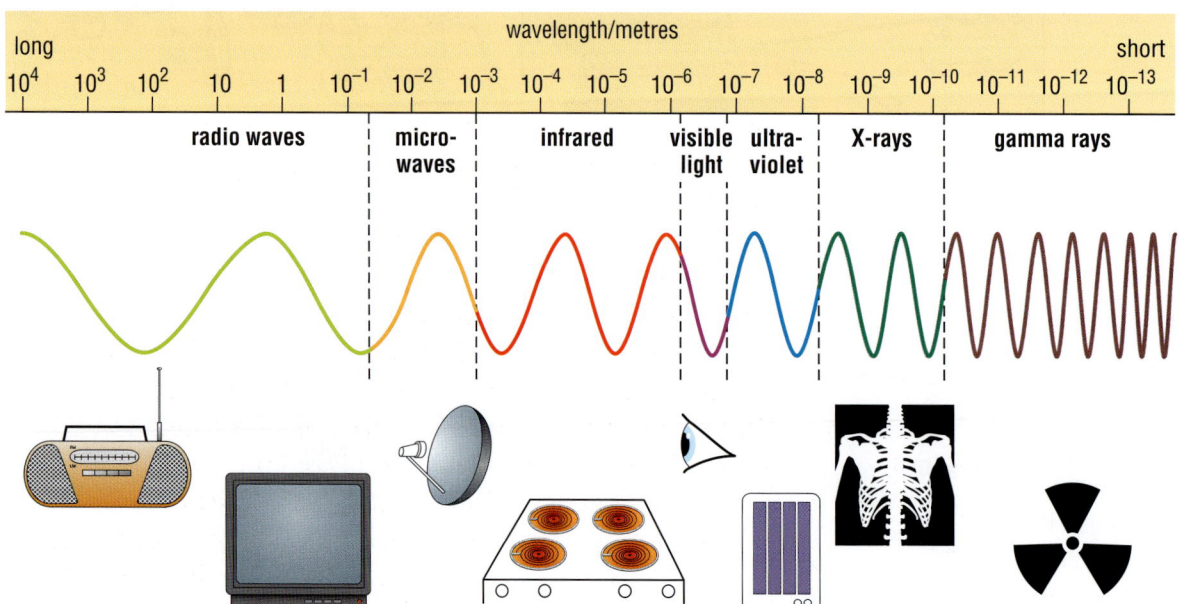

Figure 3.3 The electromagnetic spectrum.

1 Which radiation energy has
  a) the longest waves,
  b) the shortest waves?
2 Which radiation energy can our eyes detect?

Light energy

This is energy that we detect with our eyes. The light energy escaping from the Sun can be spread out by a prism or a shower of raindrops into light of different wavelengths. This forms the colours of the rainbow (see page 77) because our eyes see different wavelengths of light as different colours.

Sound energy

Sound energy is produced by the vibration of an object such as the twang of a guitar string. The energy passes though the air by the movement of the atoms and molecules. They move backwards and forwards in an orderly way. This makes a wave that spreads out in all directions from the point of the vibration. Sound energy can also pass through solids, liquids and other gases. The atoms move in a similar way to the turns on a slinky spring when a 'push-pull' wave moves along it.

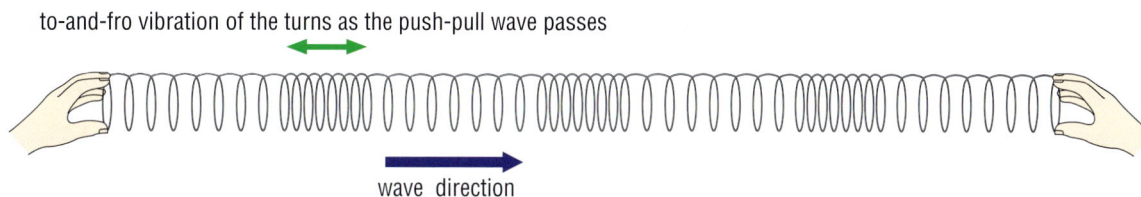

to-and-fro vibration of the turns as the push-pull wave passes

wave direction

Figure 3.4 A slinky shows how sound waves move.

3 In what ways is electrical energy put to work in your home?

Electrical energy

Electric current is the movement of electric charges through a conductor such as copper or graphite. The electric charges are given electrical energy by the battery and carry it to the working parts of a circuit. This may be a lamp, for example, where the energy is changed into light and heat.

Chemical energy

Energy can be stored in the chemicals from which a material is made. The chemicals are made from atoms that are linked together to make molecules. The chemical energy is stored in the links between the atoms. Food, fuel and the chemicals in an electrical cell (or battery) are examples of stored chemical energy.

The energy is released when the links between some of the atoms are broken and the molecule in which the energy was stored is broken down into smaller

Figure 3.5 Energy is stored in all of these objects.

molecules. For example, carbohydrates are a store of chemical energy in food. During respiration carbohydrate is broken down into carbon dioxide and water. The energy that is released in this process is used by your body to keep you alive (see *Checkpoint Biology*, page 14). The energy released by a fuel is used to heat homes, to heat water to produce steam in power stations for generating electricity (see page 98) and for the production of new materials (see *Checkpoint Chemistry*, page 126).

Kinetic energy

Any moving object has kinetic energy. The object may be as large as a planet or as small as an atom and because of its motion it can do work. When an object with kinetic energy strikes another object, a force acts on them both that will distort or set the second object moving. For example, if you move your foot and kick a stationary ball the ball moves away.

4 Look out of a window and make a list of everything you can see that has kinetic energy.

Figure 3.6 The kinetic energy of the demolition ball is transferred to the building and breaks up its structure.

5 If you are holding this book or it is resting on a table or desk why does it possess potential energy?

6 If you held a stone over the mouth of a well then let it go what would happen to the stone? Explain your answer.

Stored energy

The force of gravity between an object and the Earth pulls the object towards the centre of the planet. If an object is in a position above the surface of the Earth, it possesses stored energy called potential energy. Examples of objects with this stored energy are plates on a table, books on a shelf, a child at the top of a slide and an apple growing on a branch. Each of these objects is supported by something but if the support is removed they will accelerate to the Earth's surface and their potential energy will be released and changed into other forms.

Figure 3.7 When the objects fall their stored potential energy is released.

Some materials can be easily squashed, stretched or bent, but spring back into shape once the force acting on them is removed. They are called elastic materials. When their shape is changed by squashing, stretching or bending they store energy, which will allow them to return to their original shape.

A spring stores energy when it is stretched or squashed. Gases store strain energy in them when they are squashed. For example, when the gas used in an aerosol is squashed into a can it stores strain energy. Some of this is used up when the nozzle is pressed down and some of the gas is released in the spray (see page 153).

7 Look at Figure 3.8 on the next page. When is elastic potential energy stored and when is it released in
a) a toy glider launcher,
b) the elastic cords or springs beneath a sun lounger, and
c) a diving board?

Figure 3.8 Places where strain energy can be stored and released.

Energy changes

8 Say what main energy change takes place in the following examples:
  a) clockwork toy
  b) boy kicking a football
  c) boiling kettle on a gas ring
  d) person walking upstairs.

We use energy in many ways, for example, to cook food, light our homes and move cars and buses. When energy is used it always changes from one form to another and some always changes into heat energy. For example, when you switch on a light, electrical energy is changed to light energy and heat energy; when you play a guitar, chemical energy in your body is changed to movement energy and sound energy.

Figure 3.9 Energy changes occur when a guitar is played.

Wasted energy

9 What is the wasted energy in the energy transfers in question 8 above?

When we turn on a lamp it is because the light is useful to us. We do not use the heat that is produced so it is wasted. Sometimes that wasted energy can cause problems. For example, some machines make so much noise (wasted sound energy) that people using them have to wear ear protection (Figure 3.10).

Figure 3.10 Protection needed from noise energy.

Fuels

Many substances are burned to release their chemical energy to provide heat and light. They are called fuels. Wood, coal, gas, charcoal, oil, diesel oil, petrol, natural gas and wax are examples of fuels. The heat may be used to warm buildings, cook meals, make chemicals in industry, expand gases in vehicle engines and turn water into steam to generate electricity. Some gases and waxes are used to provide light in caravans and tents. These fuels were all formed from plants and animals that lived millions of years ago so they are known as fossil fuels.

Figure 3.11 In Nepal, ovens are often fuelled by wood.

Fossil fuels

Coal is formed from large plants which grew in swamps about 275 million years ago. These plants used energy from sunlight in the same way that plants do today.

When they died they fell into the swamps. There was a lack of oxygen in the swamp water which prevented bacteria growing and decomposing the dead plants.

Eventually the plants formed peat. Later the peat became buried and was squashed by the rocks that formed above it. The increase in pressure squeezed the water out of the peat and warmed it. These processes slowly changed the peat into coal.

Tiny plants and animals live in the upper waters of the oceans and form the plankton. When they die they sink to the ocean floor. Over 200 million years ago the dead plankton which collected on the ocean floor did not decompose because there was not enough oxygen there to allow bacterial decomposers to grow. The remains formed a layer which eventually became covered by rock. The weight of the rock squeezed the layer and heated it. This slowly converted the layer of dead plankton into oil and methane gas. This is the gas that is supplied to homes as natural gas. Several fuels are obtained from oil (see *Checkpoint Chemistry,* page 106).

Unfortunately, the supplies of fossil fuels are limited and there will come a time when there are not enough to meet our needs. As a result, scientists are trying to develop alternative sources of energy including wind power, solar power, geothermal energy and hydroelectricity (see Chapter 15).

10 What conditions helped fossil fuels to form?

♦ SUMMARY ♦

- ♦ Energy is the property of something that makes it able to exert a force and do work (*see page 29*).
- ♦ When a substance receives heat energy the particles it is made from move faster (*see page 30*).
- ♦ Radiation energy is transferred by electromagnetic waves (*see page 30*).
- ♦ Sound energy is transferred by waves in which atoms move backwards and forwards (*see page 31*).
- ♦ A battery gives electrical energy to electrical charges, which allows them to move through a conductor as a current of electricity (*see page 31*).
- ♦ Chemical energy is energy that is stored in the chemicals from which a material is made (*see page 31*).
- ♦ Movement energy is known as kinetic energy (*see page 32*).
- ♦ Some stored energy is called potential energy (*see page 33*).
- ♦ Energy can change from one form to another (*see page 34*).
- ♦ Some energy is wasted when it changes (*see page 34*).
- ♦ Substances that are burnt to release their chemical energy are called fuels (*see page 35*).

End of chapter questions

1 A group of pupils was investigating the potential energy in a 15 cm long nail. They suspended it above a block of soft clay, measured the distance to its tip then let it go. The pupils measured the depth to which the nail sank in the clay. The table shows their results for four experiments.

Height of nail above clay/cm	Depth of indent/cm
25	0.9
50	1.6
75	2.3
100	3.0

a) How do you think they measured the depth of the indent in the clay?

b) Plot a graph of their results.

c) How could you use the graph to predict the indent made by the nail from a height greater than 1 metre? Give an example.

2 A second group of pupils investigated the potential energy of a brass sphere which was dropped from different heights into soft clay. They measured the diameter of the indent made by the sphere. The table shows their results for four experiments.

Height of sphere above clay/cm	Diameter of indent/cm
5	1.0
20	1.7
50	2.2
70	2.5

a) How do you think the pupils measured the diameter of the indent?

b) Plot a graph of their results.

c) How do these results compare with the results of the first experiment?

d) Suggest a reason for any differences you describe.

e) Can the graph be used to predict indentations produced by falls from any height greater than 70 cm? Explain your answer.

4 Electrical circuits

Current electricity

Circuits

If you set up this equipment and close the switch, the lamp comes on.

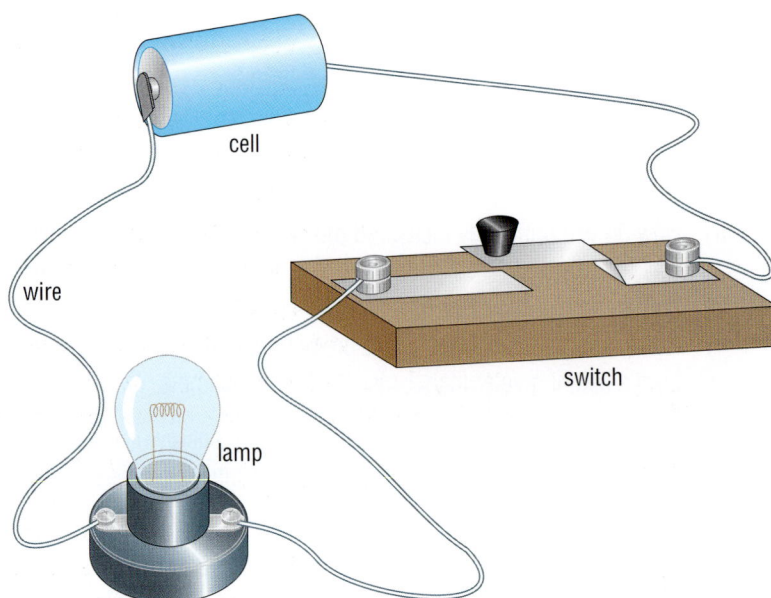

Figure 4.1 A simple circuit.

The wires of the circuit are composed of atoms that are held tightly together but around them are many electrons that are free to move. The metal filament in the lamp and the metal parts of the switch also have free electrons. When the switch is closed, the wires on either side of the switch are linked by metal contacts and a path is made along which the electrons can flow. When you open the switch, the lamp goes out. The path is broken and the electrons cannot flow.

The energy to move the electrons comes from the cell. The chemical reactions that take place in the cell make the electrons leave the cell at the negative terminal when the circuit is completed. They push their way into the wire and move the other electrons along, creating a flow or current of electricity. At the positive terminal electrons are drawn back inside the cell. The wire in the lamp filament is more resistant (see page 157) to the flow of electrons than the other wires in the circuit. As the current moves through the filament some of its electrical energy is transferred to heat energy and light energy.

1 Describe the path of an electron round the circuit in Figure 4.1 when the switch is pressed down.
2 **a)** How does the wire in the filament behave differently to other wires in the circuit when the current flows?
  b) What property of the wire accounts for this difference?

In time the chemicals which take part in the reaction inside the cell are used up. They can no longer release energy to make the electrons move and the current stops. The number of electrons in the circuit does not change; it is the chemical energy released by the cell that changes.

When Benjamin Franklin (see page 124) described substances as having positive or negative electric charges he thought that electricity flowed from a positively charged substance to a negatively charged one. His idea was taken up by other scientists until it was discovered that it was the flow of negatively charged electrons that produced a current. Franklin's idea is still used today, however; it is known as the conventional current direction.

When circuits are drawn, symbols are used for the parts or components. The use of symbols instead of drawings makes diagrams of circuits quicker to make and the connections between the components are easier to see. The symbols have been standardised like the SI units described on page 8 and are recognised by scientists throughout the world. The circuit in Figure 4.1 is shown as a circuit diagram using symbols in Figure 4.2a. The components for the circuit are the wires, cell, lamp and switch (Figure 4.2b).

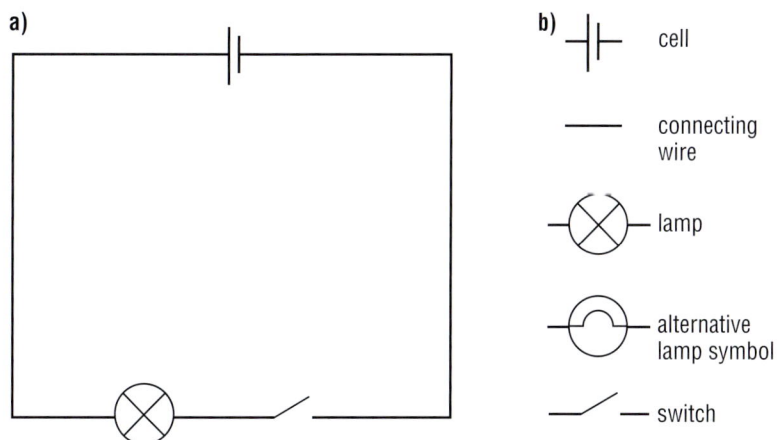

Figure 4.2 **a)** A circuit diagram and **b)** the symbols used.

3 In Figure 4.1 (page 38), the base of the cell (on the right) is the negative terminal and the cap (on the left) is the positive terminal. How can you distinguish between the positive and negative terminals in a cell in a circuit diagram?

In everyday life, cells are almost always called batteries but this is scientifically incorrect. In science a battery is made of two or more cells joined together. The symbols for a battery of two cells, three cells or more cells are shown in Figure 4.3a–c.

a)   b)   c)

Figure 4.3   a) Two cells, **b)** three cells and **c)** any number of cells.

The lining up of cells next to each other, in a row, end to end as shown in Figure 4.3, is described as arranging them in series. Other electrical components can also be arranged in series, as will be seen later.

A solid can be tested to find out if it conducts electricity by using a circuit like the one shown in Figure 4.4. The solid to be tested is secured between the pair of crocodile clips and the switch is closed. The lamp lights if the solid conducts electricity. By using the circuit, metals and the non-metal carbon, in the form of graphite, are found to be conductors of electricity. Other non-metals such as sulphur and solid compounds such as sodium chloride do not. They are insulators. Susbstances such as wood and plastic are also insulators.

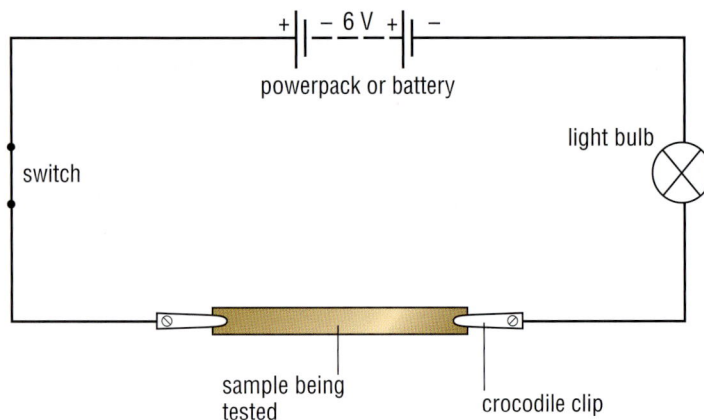

Figure 4.4   A circuit for testing conduction of solid materials.

Other circuit components

Resistors

The property of resistance will be discussed in detail on page 157. If a short piece of wire which has a high resistance is included in a simple circuit with a cell, a switch and a lamp, the lamp will shine less brightly than before. If a longer length of high resistance wire is included in the circuit the light will shine even more dimly.

A component that is designed to introduce a particular resistance into a circuit is called a resistor.

4 How does the length of a high resistance wire affect the flow of current through the circuit?

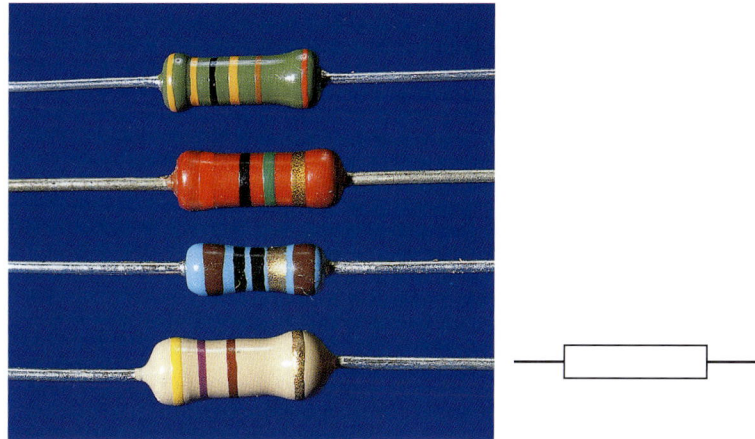

Figure 4.5 Four resistors and the symbol for a resistor.

A variable resistor can be made in which a contact moves along the surface of a resistance wire and brings different lengths of the wire into the circuit. This device is sometimes called a rheostat. In order to make it more compact, the length of the wire is wound in a coil and the contact is made to move freely across the top of the coil.

5 In Figure 4.6, which way should the contact be moved to
a) increase,
b) decrease
the resistance in the part of the wire included in the circuit between A and B?

6 Figure 4.7 shows a variable resistor in a dimmer switch. How would you turn the switch to make the lights
a) brighter,
b) dimmer?
Explain your answer.

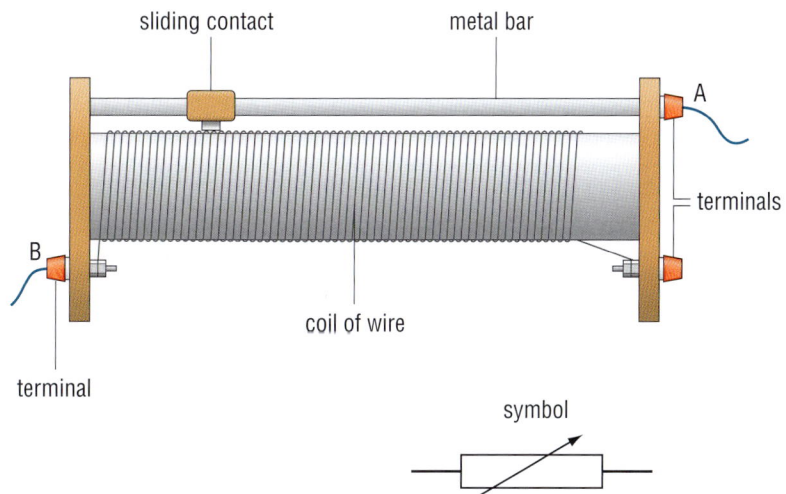

Figure 4.6 A variable resistor and its symbol.

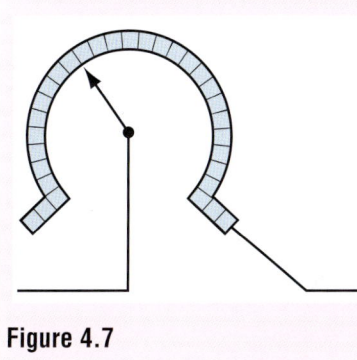

Figure 4.7

In Figure 4.6 the current passes through terminal A, along the bar, through the sliding contact and coil of wire to terminal B. When the contact is placed on the far left the current passes through only a few coils of the wire. As the contact is moved to the right the current flows through more of the wire and encounters greater resistance. When the contact is moved from the right to the left the current flows through fewer coils of the wire and encounters less resistance.

7 Draw a circuit diagram for
circuits that have:
   a) 2 cells, a variable resistor,
      2 lamps and a switch.
   b) 3 cells, a lamp, a buzzer and
      a switch.
   c) 2 cells, a variable resistor, a
      buzzer and a switch.

Buzzers

A buzzer is an electrical device in which one part vibrates strongly when a current of electricity passes through it. The vibrations produce the sound (see also page 80).

Figure 4.8   A buzzer and its symbol.

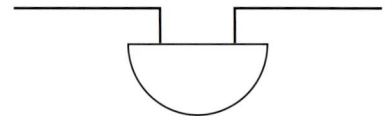

Fuses and circuit breakers

When a current flows through a circuit some energy is lost as heat (thermal) energy. If the size of the current increases, the amount of heat (thermal) energy released also increases. A fuse is a device that contains a wire which melts when the current flowing through it reaches a certain value. When the wire melts, it breaks and so also breaks the circuit and stops the current flowing.

Electrical appliances are designed to work when a current of a certain size flows through them. If the current is too large, the appliance may be damaged. An unusually large current can occur in a household circuit in two ways. It may occur when the insulation in a cable is worn and the wires in the cable touch each other. This causes a short circuit. It may also occur if too many appliances are plugged into one socket.

Fuses are used to stop the flow of a current when it becomes too large for the circuit. They may be present in plugs (see Figure A) and/or in the appliances themselves. In the past each circuit in a home was protected by a fuse in a consumer unit or 'fuse box'. Today circuit breakers are used instead of fuses in consumer units. A circuit breaker is a switch that is sensitive to the size of the current flowing through it. If the current is too large, the switch opens and breaks the circuit. The switch can be closed and the circuit used again once the cause of the problem has been identified and corrected.

1 How could you prevent a fire in
the home being caused by an
electrical fault?

2 Why should you always switch
off a circuit before replacing a
fuse?

(continued)

a) A three amp fuse.

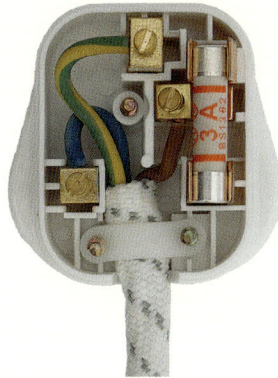

b) The inside of a plug with no fuse.

c) The inside of a plug with a fuse in place.

Figure A

♦ SUMMARY ♦

♦ A closed or completed circuit is needed for an electrical current to flow (*see page 38*).

♦ When circuits are drawn, symbols are used for the parts or components (*see page 39*).

♦ The lining up of cells in a row is called arranging them in series (*see page 40*).

♦ Materials which allow a current of electricity to pass through them are called conductors (*see page 40*).

♦ Materials which do not allow a current of electricity to pass through them are called insulators (*see page 40*).

♦ Resistors control the amount of electricity passing through a circuit (*see page 40*).

♦ A buzzer vibrates to produce sound when a current of electricity passes through it (*see page 42*).

♦ Fuses and circuit breakers prevent circuits carrying too much electricity (*see page 42*).

End of chapter question

1 Make a list of the components in each of these three circuits.

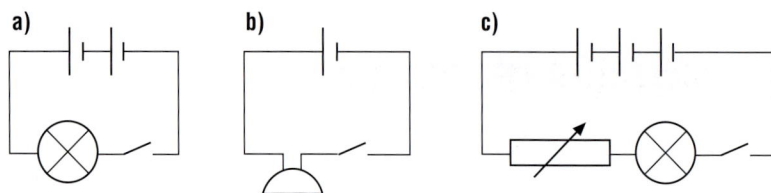

Figure 4.9

5 Particle model

In our world matter exists in three states – solid, liquid and gas.

Particles of matter

Observations on the three states of matter and how they change can be explained by considering that matter is made of particles. This is called the 'particle theory of matter'.

Particles in the three states of matter

In solids, strong forces hold the particles together in a three-dimensional structure. In many solids the particles form an orderly arrangement called a lattice. The particles in all solids move a little. They do not change position but vibrate to and fro about one position.

In liquids, the forces that hold the particles together are weaker than in solids. The particles in a liquid can change position by sliding over each other.

In gases, the forces of attraction between the particles are very small and the particles can move away from each other and travel in all directions. When they hit each other or the surface of their container they bounce and change direction.

1 According to the particle theory, why do liquids flow but solids do not?
2 How is the movement of particles in gases different from the movement of particles in liquids?

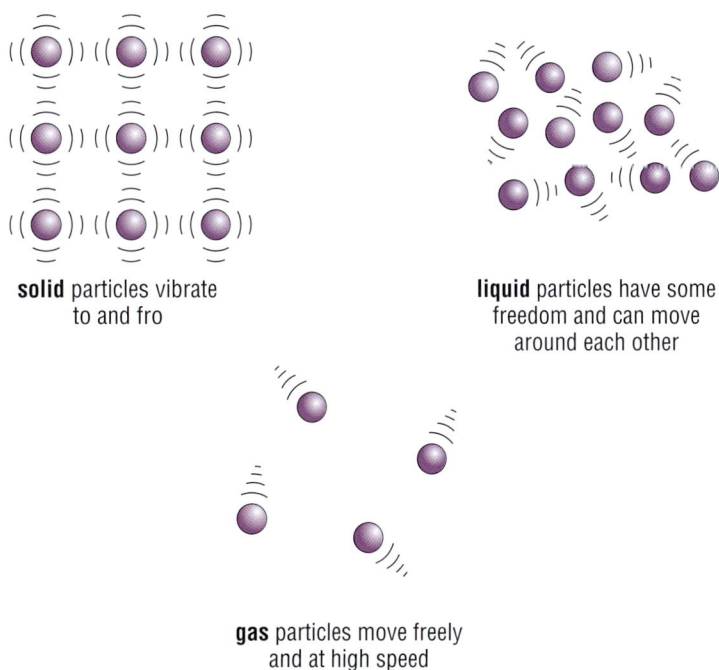

solid particles vibrate to and fro

liquid particles have some freedom and can move around each other

gas particles move freely and at high speed

Figure 5.1 Arrangement of particles in a solid, a liquid and a gas.

How materials change with temperature

Changes in solids

When a solid is heated it expands and when it cools it contracts. For example, when it is cool a metal bar fits inside a metal gauge (Figure 5.2). Its end will also fit through the hole in the gauge. When the bar has been heated it will no longer fit in the gauge or in the hole. When the bar cools down it will fit in the gauge and hole again.

3 In which ways did the metal bar expand?

4 If the gauge was heated instead of the bar, would the length of the bar fit in the gauge? Explain your answer.

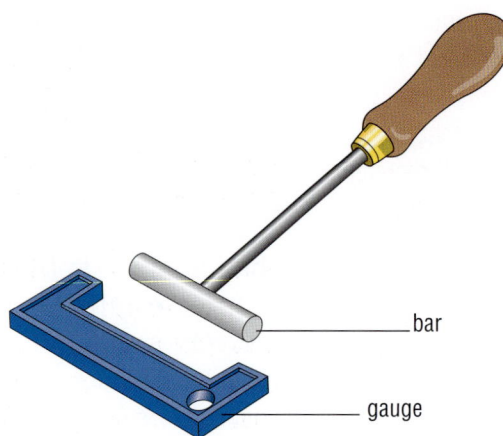

bar

gauge

Figure 5.2 This apparatus shows that metals expand when heated.

The expansion of materials due to heating must be taken into account whenever the materials are likely to encounter changes in temperature. For example, metal pipes carrying hot water in large central heating systems are connected together by expansion joints (Figure 5.3) which allow the pipes to lengthen without pushing into each other.

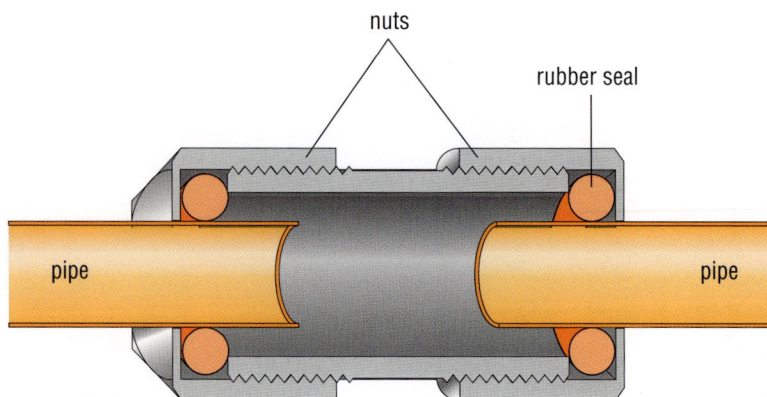

nuts

rubber seal

pipe

pipe

Figure 5.3 An expansion joint.

The temperature changes due to the weather can also cause expansion and contraction. The end of a bridge, for example, may be set on rollers so that as it lengthens it simply moves over its support and does not push into the adjoining roadway.

Figure 5.4 Rollers in the wall of a foot bridge prevent any damage when the bridge expands in hot weather.

The lines on the first railways were made of short lengths of rail that had gaps between their ends. These allowed the rails to expand in hot weather without buckling. When trains passed over the junctions they made a clickety-click sound. Today the lengths of metal used in railways lines are much longer than those used in the past and a very large gap would be needed if they were to be joined together in the same way. The longer rails have tapered ends which overlap each other (Figure 5.5). This arrangement allows the ends to slide past each other when the metal expands, without affecting the safety of the train running along them.

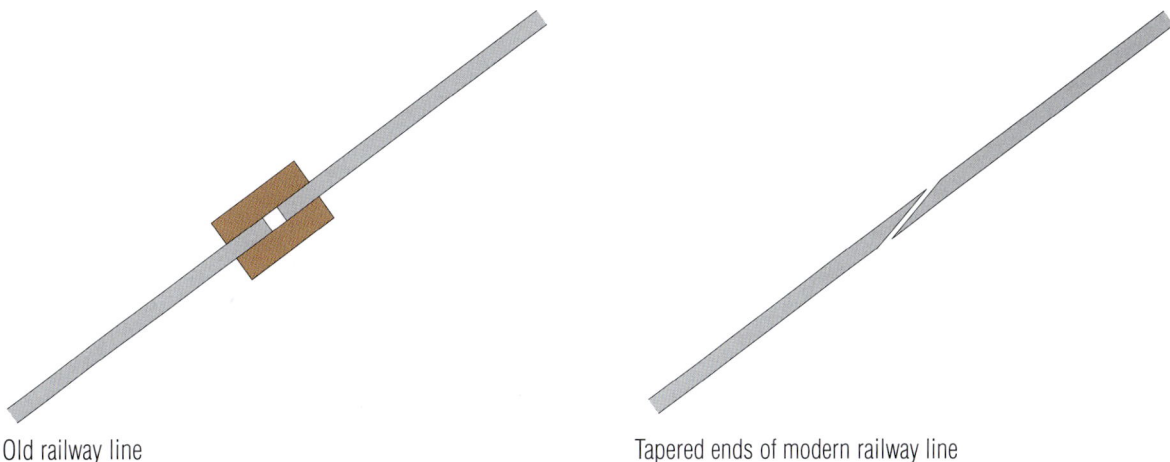

Old railway line

Tapered ends of modern railway line

Figure 5.5

5 Figure 5.6 shows an overhead power cable suspended from pylons in winter. Draw how it might appear in summer. Explain your answer.

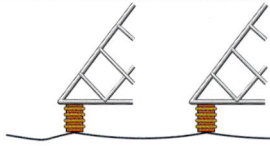

Figure 5.6

6 Metal expands more than glass when both are heated. How can this fact be used to help open a metal screw top on a glass bottle which appears to be stuck?

7 Why will the fire alarm in Figure 5.8a ring when there is a fire?
8 a) How does turning the screw down on the iron in Figure 5.8b alter the temperature at which the circuit is broken?
b) When the screw is set at the bottom the iron can be used on linen fabrics. When the screw is set at the top the iron can be used for nylon fabrics. Which fabric needs most heat?

Different solids expand by different amounts for a given temperature rise. For example, brass expands more than iron. When strips of these two metals are stuck together they form a bimetallic strip. If the bimetallic strip is heated it bends because the length of the brass strip becomes greater than the length of the iron strip, as Figure 5.7 shows.

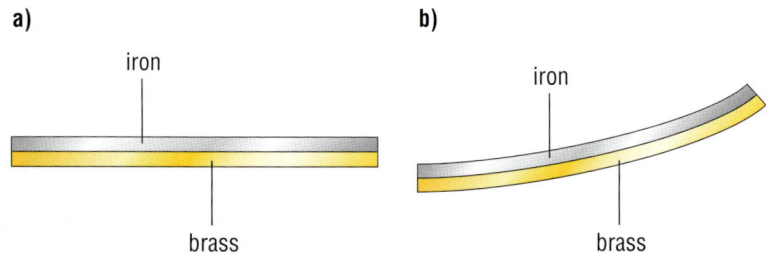

Figure 5.7 A bimetallic strip **a)** when cold and **b)** when hot.

A bimetallic strip may be used in a fire alarm and as a thermostat (temperature regulator) in an electric iron. The bending of the metal as the temperature rises makes or breaks the electrical circuit.

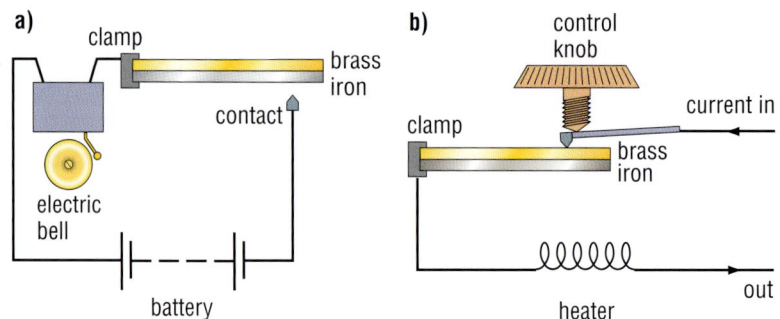

Figure 5.8 **a)** A fire alarm circuit and **b)** a circuit in an electric iron.

Changes in liquids

When liquids are heated they generally expand much more than solids for a given temperature rise. They also contract to their original volume when they are cooled. The expansion and contraction of a liquid can be demonstrated by setting up the apparatus in Figure 5.9. When the water gets hot it expands and rises up the glass tube. When the water is allowed to cool the level of water in the tube falls.

9 How could you use the apparatus in Figure 5.9 to compare the expansion of different liquids? What precautions would you have to take when testing flammable liquids such as alcohol and paraffin?

Figure 5.9 Demonstrating the expansion of water.

However, water has a strange property – as it is cooled it expands again when its temperature drops below 4 °C and continues to expand until it reaches 0 °C (see *When water is cooled* below).

When water is cooled

Almost all liquids contract when they are cooled and when they freeze they contract further to form a denser solid. When water cools down it initially contracts like other liquids, but this changes when it reaches 4 °C. As it cools from 4 °C to 0 °C it expands. Water below 4 °C is, therefore, less dense than water at 4 °C. When water has a range of temperatures close to 0 °C, the colder water rises above the warmer water and collects at the surface. If the cooling of the water continues the water at the surface turns to ice at 0 °C and expands even more. The ice is, therefore, less dense than the water below it so the ice floats on the surface of the cold water.

In very cold winter weather this means that the water in a lake or pond freezes from the surface downwards because the coldest water is at the surface. Fish can remain alive and active in the warmer, denser water at the bottom of the pond while the surface is frozen (see Figure 5.10 overleaf).

10 a) If water behaved like other liquids, how would a pond freeze in winter?
  b) How would fish be affected by the way the pond froze?
11 Explain what has happened to the milk in Figure 5.11.

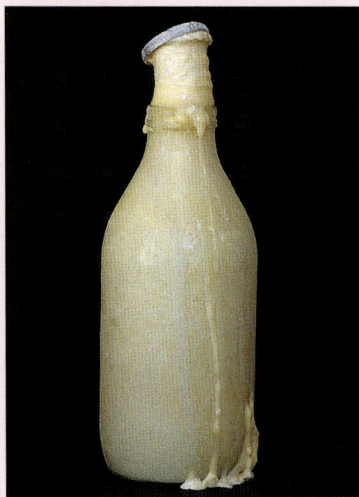

Figure 5.11

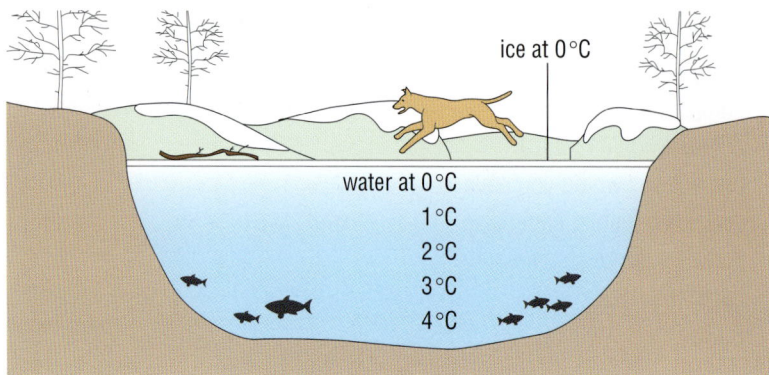

Figure 5.10 Section through a pond in winter.

The cold water in pipes in a building may freeze in very cold winter weather if they are not insulated. The water expands as it freezes, pushing on the walls of the pipes and bursting them. You do not find out that this has happened until the ice thaws again.

Changes in gases

Like most liquids and solids, gases expand when they are heated and contract when they are cooled. If a gas and a liquid undergo the same temperature rise the gas expands about ten times more than the liquid.

Hot air balloons

When the air in a hot air balloon is heated it expands and fills the balloon. The density of the air in the balloon falls. It becomes less dense than the air around it. The surrounding air then exerts an upthrust force which is greater than the combined weight of the balloon, basket, pilot and passengers. The density of the surrounding air decreases with an increase in height. The balloon will continue to rise until it reaches air that is sufficiently less dense for the weight and upthrust to balance. If the air in the balloon is allowed to cool it contracts and its density increases. The upthrust is now less than the weight so the balloon sinks until it reaches air that has the same density. The pilot must heat the air with the gas burners to make the balloon rise again.

Figure 5.12 Hot air balloons make use of the fact that air expands when heated.

Explosions

An explosion occurs when a gas is made to expand very strongly and very quickly. This creates large forces that push on everything around. Explosives are used in quarries – the force of the expansion is strong enough to break up rocks.

Figure 5.13   Explosives are breaking up the rocks in this quarry.

Inside a car engine

A car engine has cylinders in which small explosions occur. A mixture of air and petrol vapour is ignited by a spark plug, and the hot gases, which expand quickly, push down a piston. The downward force of the piston is changed into a turning force by other parts of the engine, and this is used to turn the car's wheels.

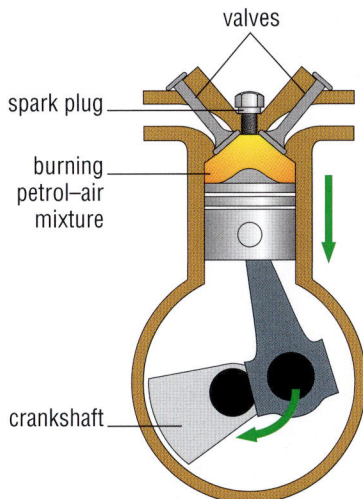

valves

spark plug

burning
petrol–air
mixture

crankshaft

Figure 5.14   A cylinder in a car engine.

12 How does the size and speed of expansion of a gas compare with the size and speed of expansion of a solid?

Melting and boiling

Figure 5.1 shows the arrangement of particles in a solid, liquid and gas. You can also look at Figure 5.1 to help you understand how the particles in a substance behave as it is heated and changes from a solid to a liquid, and then from a liquid to a gas. In a solid the particles are held firmly together but as they get hotter, they come to vibrate so strongly that they begin to slide over each other. When this occurs, the solid has melted and become a liquid. The temperature at which a solid melts is called its melting point.

As a liquid receives more and more heat, its particles become more and more active. Eventually they become so active that they lose their grip of each other and become a gas. This can occur at the surface of the liquid or inside the liquid where the gas particles form bubbles. This process of forming bubbles is called boiling, and the temperature at which it occurs is called the boiling point of the liquid.

◆ SUMMARY ◆

- ◆ The particles in the three states of matter behave differently (*see page 45*).
- ◆ Solids expand when they are heated and contract when they are cooled (*see page 46*).
- ◆ Liquids expand when they are heated and contract when they are cooled (*see page 48*).
- ◆ Gases expand when they are heated and contract when they are cooled (*see page 50*).
- ◆ The temperature at which a solid changes into a liquid is called the melting point (*see this page*).
- ◆ The temperature at which a liquid changes into a gas is called the boiling point (*see this page*).

End of chapter question

1 If you live in a country where salt is spread on the roads in winter, you may have tried to think of the reason for this. You may have thought along these lines: Ice forms on the road. Ice makes cars skid. This is unsafe. Salt makes the roads safe in some way. Perhaps the salt affects the ice in some way. One way it could affect the ice is to melt it.

We think of substances being melted by having their temperatures raised. Perhaps the salt dissolves in the water on the ice surface and makes a solution with a lower melting point than ice. If this happens the solution will freeze at a lower temperature than the ice, so will be liquid at 0 °C.

To test this idea imagine you are given a large beaker of ice, two filter funnels, two measuring cylinders (to support the filter funnels), two thermometers, a small beaker of salt and a stopclock.

a) Set out a plan for an experiment to test whether salt makes ice melt.

b) If it does, what would you expect to happen?

6 Magnets and electromagnets

Three metallic elements show strong magnetic properties. They are iron, cobalt and nickel. Steel is a metal alloy which can show magnetic properties. It is made from iron and carbon. Steel can also be mixed with other metals to make an alloy which does not show magnetic properties. For example, stainless steel is made from steel, chromium and nickel and it does not show magnetic properties.

Materials that show magnetic properties do not show them all the time. For example, steel paper clips do not generally attract and repel each other. When a material is showing magnetic properties it is said to be magnetised and is known as a magnet. The most widely used magnets used to be made from steel but most magnets are now made from mixtures of the magnetic metals. Alnico is an example.

It is thought that the word 'magnet' comes from the name of the ancient country of Magnesia which is now part of Turkey. In this region large numbers of black stones were found which had the power to draw pieces of iron to them. The black stone became known as lodestone or leading stone because of the way it could be used to find directions (see *Early discoveries about magnetism*, page 60). Today it is known as the mineral magnetite and it has been found in many countries.

1 Which three metals do you think might be present in Alnico? Explain your answer. Which ones are magnetic?

Figure 6.1 Magnets are not only used to hold messages on fridge doors but a magnetic strip in the fridge door is used to hold it closed.

Figure 6.2 Magnetite is a naturally occurring magnet.

2 How do magnetic materials differ from non-magnetic materials in
  a) what they are made from, and
  b) their properties?

The behaviour of magnets

Magnets can attract or repel other magnets and can attract any magnetic material even if it is not magnetised. When suspended from a thread, a bar magnet aligns itself in a north–south direction.

Non-magnetic materials, such as wood, paper, plastic and most metals, cannot be magnetised and so can do none of these things. Some, such as paper and water, can let the force of magnetism pass through them while other materials, such as a steel sheet, do not let the force of magnetism pass through them.

The strength of the magnetic force

At each end of a bar magnet is a place where the magnetic force is stronger than at other places in the magnet. These places where the magnetic force is strongest are called the poles of the magnet. The end of the magnet which points towards north when the magnet is free to move is called the north-seeking pole or north pole. At the other end of the magnet is the south-seeking pole or south pole.

When the north pole of one magnet is brought close to the south pole of another magnet that is free to move, the south pole moves towards the north pole. Similarly, a north pole is attracted to a south pole. However, two south poles repel each other, as do two north poles. These observations can be summarised by the phrase 'different poles attract, similar poles repel'.

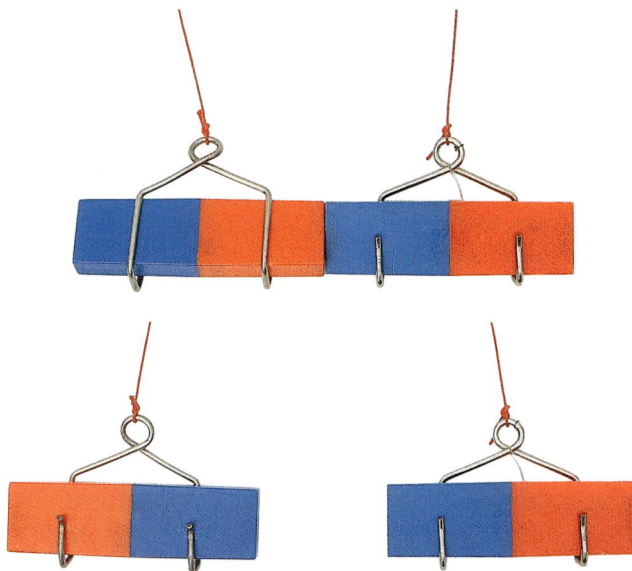

Figure 6.3 Unlike poles attract and like poles repel.

55

3 What is the relationship between the distance from the magnet and the strength of the magnetic force?

4 Why do the two lower paper clips in Figure 6.4 join together?

5 A nail is magnetised by being in contact with one end of a magnet. Can it still attract magnetic materials to it when it has left the magnet? Explain your answer.

6 If three steel paper clips are attached in line to the bottom of a magnet the lowest paper clip is attached less strongly to the middle one than the middle paper clip is attached to the top one. A fourth paper clip cannot be added. Look at your answer to question 3 and explain why the paper clips behave in this way.

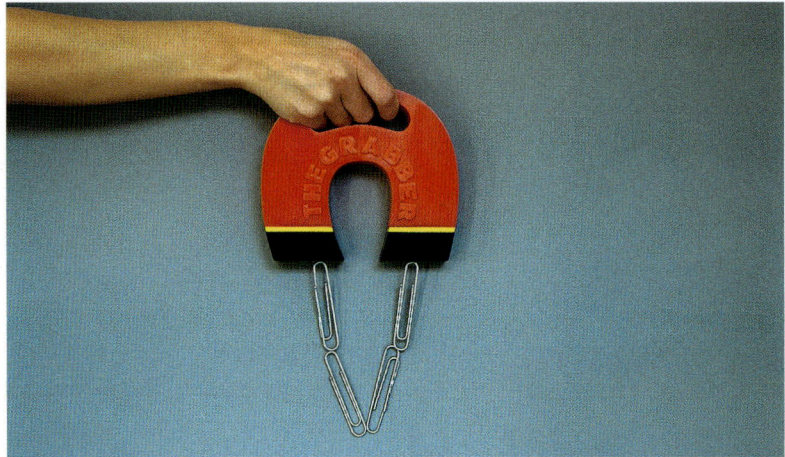

Figure 6.4 Paper clips attracted by a magnet are themselves magnetised.

If you bring a steel paper clip (which is not magnetised) towards either pole of a magnet you will feel the pull of the magnetic force become stronger as the paper clip gets closer to the pole. As you move the paper clip away again you will feel the pull of the magnet become weaker.

When a material that can show magnetic properties, such as a steel paper clip, is attracted to the end of a magnet it also becomes a magnet and can attract other magnetic materials to it. The paper clip has been made into a magnet by a process called magnetic induction. When the paper clip is moved away from the magnet it loses its magnetism.

Inside a magnet

a)

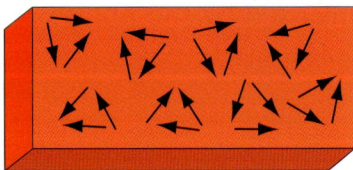

b)

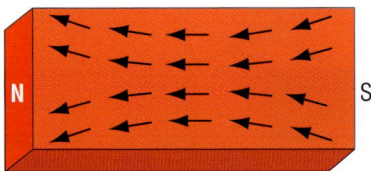

Figure 6.5 a) Domains in an unmagnetised material and **b)** domains in a magnet.

Groups of particles from which a magnetic material is made form tiny regions called domains. Each domain behaves like a tiny magnet. If the domains are arranged at random (Figure 6.5a) the material does not attract other magnetic materials to it although it can be attracted to a magnet. It also does not point north–south when it is free to move.

Magnetic domains can be made to arrange themselves in line. Then all their north poles face in one direction and all their south poles face in the opposite direction. This arrangement produces a north and a south pole in the material as a whole (Figure 6.5b). When the material is in this condition it has been magnetised – it is now a magnet.

7 If you cut a magnet in half does each half become a magnet? Explain your answer.

8 A piece of steel can be made into a magnet by repeatedly stroking it with a magnet as shown in Figure 6.6. How does this affect the domains?

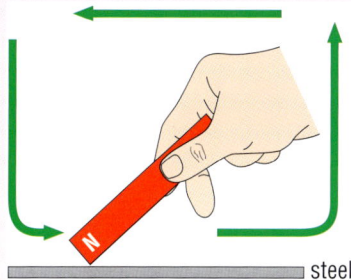

Figure 6.6

Some materials, such as steel, are magnetically 'hard' and once domains have been aligned they tend to stay put. Others, such as iron, are magnetically 'soft' and domains soon rotate again to random positions, so the material loses its magnetism.

The magnetic field

The region around a magnet in which the pull of the magnetic force acts on magnetic materials is called the magnetic field.

The field around a magnet can be shown by using a piece of card and iron filings. The card is laid over the magnet and the iron filings are sprinkled over the paper. Each iron filing has such a small mass that it can be moved by the magnetic force of the magnet if the paper is gently tapped. The iron filings line up as shown in Figure 6.7. The pattern made by the iron filings is called the magnetic field pattern.

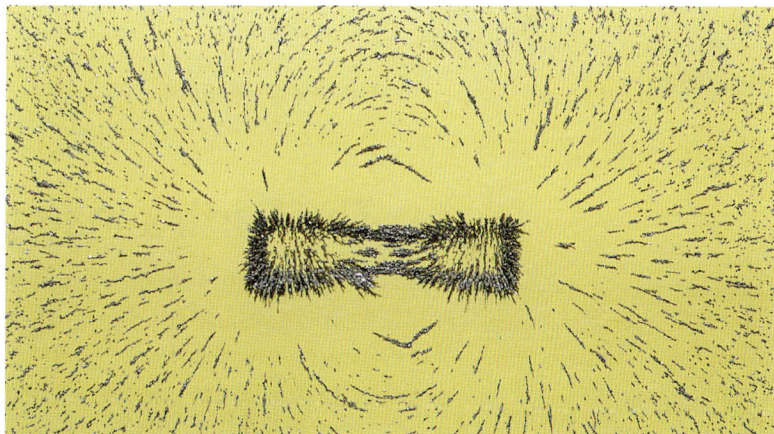

Figure 6.7 The magnetic field pattern of a bar magnet shown by iron filings.

The iron filings appear to form lines around the magnet. This phenomenon can be checked by using a plotting compass and a piece of paper and pencil. The magnet is placed in the centre of the paper and the plotting compass is placed on one side of the magnet close to its north pole. The north pole of the compass will point away from it. The position of the north pole of the compass is marked on the paper and the plotting compass is then moved so that its south pole is over the mark made on the paper. The position of the north pole is marked again with the plotting compass in the new position and the process is repeated until the plotting

9 How does the information from the activity with the plotting compass compare with the field pattern produced by the iron filings?

10 Figure 6.9 shows iron filings spread out when in contact with the end of a bar magnet. Make a drawing of how you think the field lines are arranged all around the magnet.

Figure 6.9

compass reaches the south pole of the magnet. If the points marking the positions of the north pole of the compass are joined together by a line running from the north pole to the south pole of the magnet (Figure 6.8a), this will represent one of the magnetic 'lines of force' forming the field pattern. Arrows should be drawn on the lines, pointing from the magnet's north pole to its south pole (Figure 6.8b).

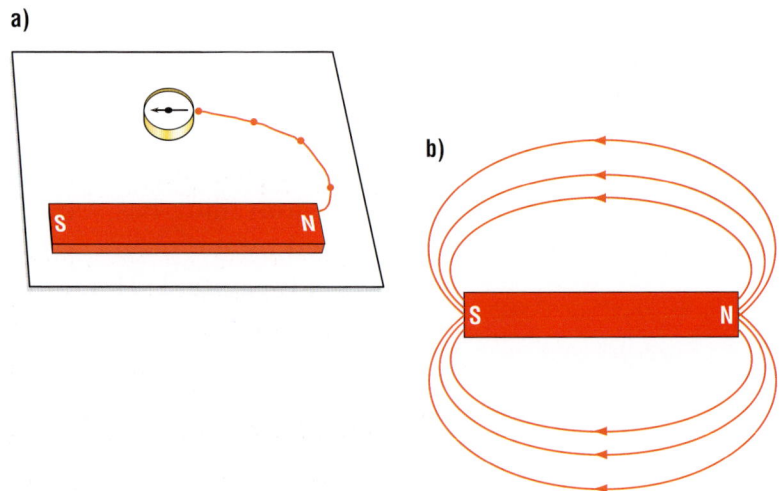

Figure 6.8 **a)** Drawing a magnetic line of force and **b)** the magnetic field pattern around a bar magnet.

The Earth's magnetic field

At the centre of the Earth is the Earth's core. It is made from iron and nickel and is divided into two parts – the inner core made of solid metal and the outer core made of liquid metal. As the Earth spins the two parts of the core move at different speeds and this is thought to generate the magnetic field around the Earth and make the Earth seem to have a large bar magnet inside it.

The Earth spins on its axis which is an imaginary line that runs through the centre of the planet. The ends of the line are called the geographic north and south poles. Their positions on the surface of the Earth are fixed. Magnetic north – towards which the free north pole of a magnet points – is not at the same place as the geographic north pole (Figure 6.10), and it changes position slightly every year.

The north magnetic pole originally got its name because it is the place to which the north poles of bar magnets point. In reality it is the Earth's south magnetic pole because it attracts the north poles of magnets.

11 a) Look at the field pattern around the Earth in Figure 6.10. Which pole of the imaginary bar magnet inside the Earth coincides with magnetic north?

  b) Draw a bar magnet inside the Earth and label its poles. Also label the position of the south magnetic pole on the Earth's surface.

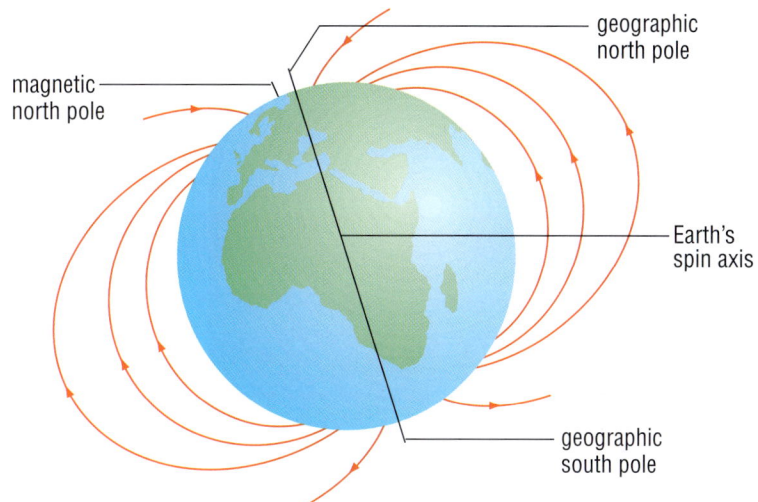

Figure 6.10   The Earth's geographic and magnetic poles do not coincide.

Similarly the south magnetic pole is really the Earth's north magnetic pole because it attracts the south poles of bar magnets. However for most purposes the old, and incorrect, names for the magnetic poles are still used.

The Sun and the Earth's magnetic field

Billions of electrically charged particles leave the Sun every second. They stream out through space and form the solar wind. Those particles that approach the Earth or pass close to it distort the Earth's magnetic field.

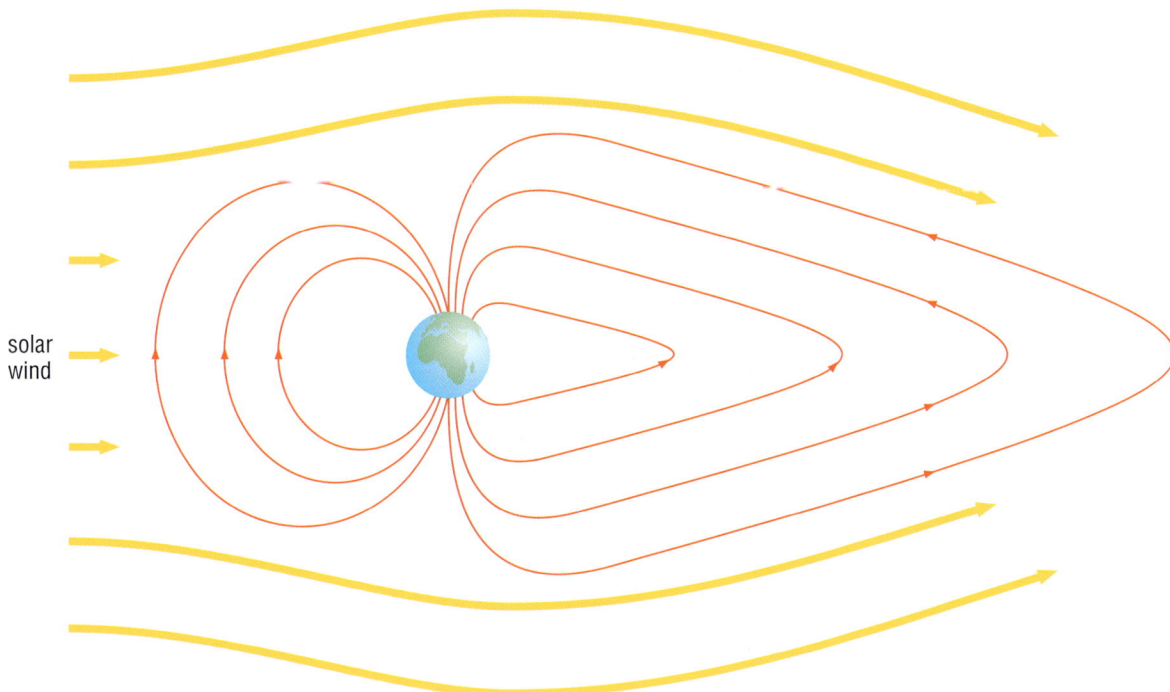

Figure 6.11   Charged particles in the solar wind distort the Earth's magnetic field.

Figure 6.12 The aurora borealis.

Some of the particles in the solar wind are drawn in towards the Earth's poles and when they enter the upper atmosphere they collide with particles of gas. Light energy is released as a result of these collisions. It appears like a shimmering curtain of colour which is known as the northern lights (aurora borealis) or the southern lights (aurora australis). The intensity of the light emitted varies with the strength of the solar wind.

Early discoveries about magnetism

The discovery of how magnetite can be used to identify the direction of north and south is thought to have occurred in China over 4500 years ago. It is thought that the discovery may have come about in the following way.

In ancient China people were concerned about where they were buried. They took advice from people who spun an object on a board to find the best direction for the grave. As the practice continued the people who used the boards and spinners charged more money and used more expensive and unusual materials, such as magnetite, to attract even higher fees for their work. When magnetite was used as a spinner it always settled in a north–south direction, while the other objects settled in any random direction. The importance of magnetite in direction finding led to the development of the compass.

1 How did magnetite become used for direction finding?
2 How is the knowledge about the compass thought to have spread?

Figure A An early Chinese compass.

The knowledge of using magnetite for direction finding is believed to have slowly passed to other countries as they traded with one another. It reached Europe some time after AD950.

Petrus de Peregrinus (also known as Peter the Pilgrim) was a French engineer who lived in the 13th century. He experimented on the way magnets could attract and repel each other and how they could point north and south. He believed that the magnet pointed to the outer sphere of the heavens. Compasses at that time were made by floating the magnetic needle on water but Peregrinus showed that attaching the needle to a pivot made the compass easier to use.

Figure B An Italian mariner's compass made in 1580, using a pivot originally designed by Peregrinus.

3 How did the work of Peregrinus help Gilbert with his scientific modelling?

4 How did Gilbert's explanation of the reason for magnets pointing north–south differ from the explanation given by Peregrinus?

William Gilbert (1544–1603) was an English scientist and doctor to Elizabeth I, who made many experiments on magnets and disproved beliefs, such as garlic destroys magnetism and rubbing a diamond on a piece of iron makes the iron into a magnet.

Gilbert suspended a magnetic needle so that it could move both horizontally and vertically and discovered that the needle also dipped as it pointed north–south. He extended his investigation by using a model of the Earth made out of a sphere of lodestone (magnetite). He put a compass with a pivot at different places on the surface of his model Earth and showed that the dip varied with the position of the compass on the sphere, just as it did with compasses at different places on the surface of the Earth. From this investigation Gilbert described the Earth as behaving as if it contained a huge magnet.

Figure C William Gilbert.

For discussion

Was the ancient Chinese practice of grave positioning a scientific process? Explain your answer.

The link between magnetism and electricity

Hans Christian Oersted (1777–1851) was a Danish physicist who studied electricity. In one of his experiments he was passing an electric current along a wire from a battery when he noticed the movement of a compass needle which had been left near the wire. This chance observation led to many discoveries of how magnetism and electricity are linked together, and many modern applications.

When an electric current passes through a wire it generates a magnetic field around the wire. A compass can be placed at different positions on a card around the wire, as shown in Figure 6.13, and lines of force can be plotted.

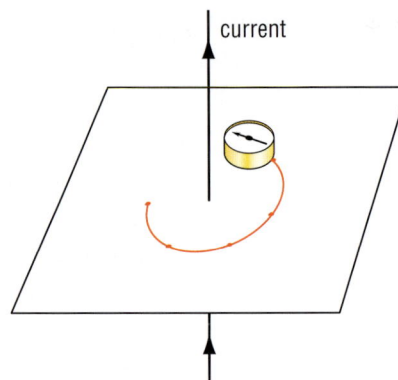

Figure 6.13 Plotting magnetic field lines around a current-carrying wire.

When the current flows up through the card, the field shown in Figure 6.14a is produced. When the current flows down through the card, the field shown in Figure 6.14b is produced.

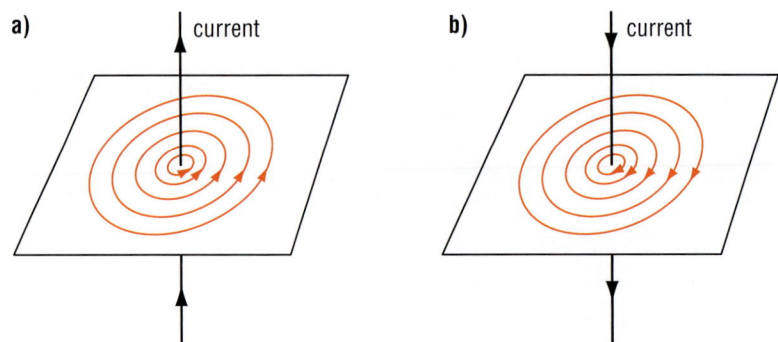

Figure 6.14 The magnetic field around a current-carrying wire.

Lines of force on diagrams of magnetic fields show not only the direction of the field as given by a plotting compass but also the strength of the field in different places. The lines of force are close together where the field is strong and wider apart where the field is weaker.

If the wire is made into a coil and connected into a circuit, a magnetic field is produced around the coil as shown in Figure 6.15.

12 How are the fields in Figure 6.14a and b different?

13 How does the strength of the magnetic field around the wire vary?

14 Compare the magnetic field of a bar magnet (Figure 6.8b) with that produced by a current in a wire coil (Figure 6.15).

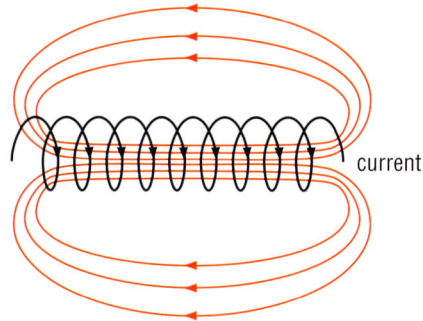

current

Figure 6.15   The magnetic field around a current-carrying coil.

If a piece of steel is placed inside the wire coil and the current is switched on, the magnetism of the coil and the steel is stronger than that of the coil alone. The current flowing through the coil induces magnetism in the steel. When the current is switched off the steel keeps some of the magnetism it acquired because it is magnetically hard (see page 57).

If a piece of iron is placed inside the coil it makes an even stronger magnet when the current is switched on than the steel did. When the current is switched off the iron loses its magnetism completely because it is magnetically soft (see page 57).

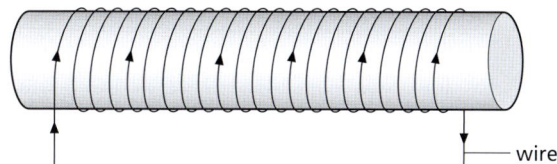

wire

Figure 6.16   Steel bar within a coil.

The electromagnet

An electromagnet is made from a coil of wire surrounding a piece of iron. When a current flows through the coil magnetism is induced in the iron, and the coil and iron form a strong electromagnet. When the current is switched off the electromagnet loses its magnetism completely, straight away. This device, which can instantly become a magnet then instantly lose its magnetism, has many uses. For example, a large electromagnet is used in a scrapyard to move the steel bodies of cars (Figure 6.17).

15 Describe how you think an electromagnet can be used to make a stack of scrapped cars three cars high.

Figure 6.17   An electromagnet in use in a scrapyard.

The reed switch

A reed switch (Figure 6.18) is a magnetic switch. It has two pieces of soft iron, called the reeds, supported by metal which has a springy property. The reeds are enclosed in a glass container which is filled with an inert gas. This gas is used instead of air because the metal does not react with it and so does not corrode.

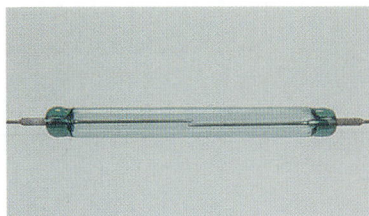

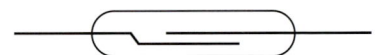

Figure 6.18   A reed switch and its symbol.

When a magnet is brought close to the reed switch it makes the soft iron reeds become magnets (Figure 6.19). The opposite poles on the free ends of the reeds attract each other. The magnetic force between them bends the two pieces of springy metal so that the two reeds touch and close the circuit, allowing the current to flow. When the magnet is taken away the soft iron reeds lose their magnetism and the tension force in the springy metal pulls the reeds apart.

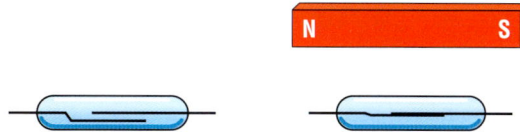

Figure 6.19 How a reed switch works.

16 How would metal corrosion affect a reed switch? Explain your answer.

17 If a door is fitted with a burglar alarm as in Figure 6.20, how is the alarm set off when the door is opened?

This type of reed switch is used in burglar alarms where the reed switch is placed in the door frame and magnets are placed in the door and the door frame.

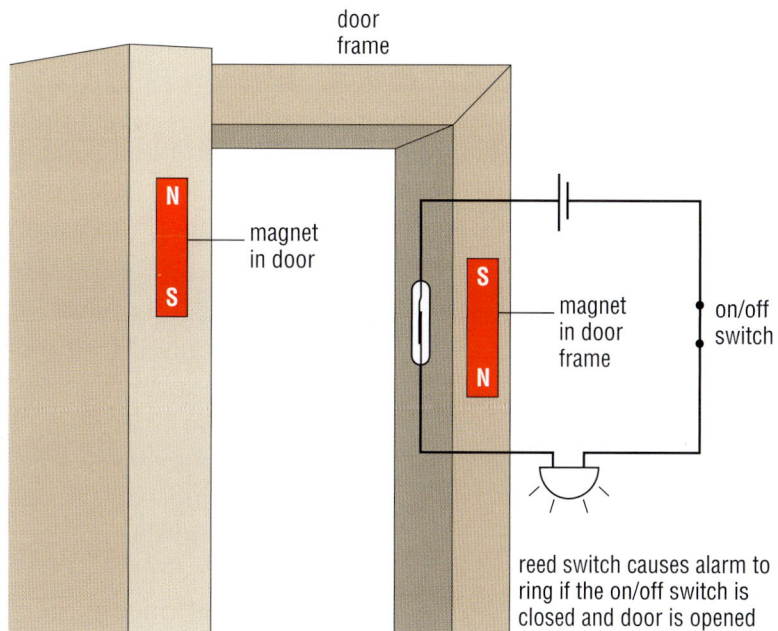

Figure 6.20 A reed-switch operated burglar alarm.

The electric bell

Look at Figure 6.21 and see if you can work out the path the current takes through the circuit when the switch is pushed.

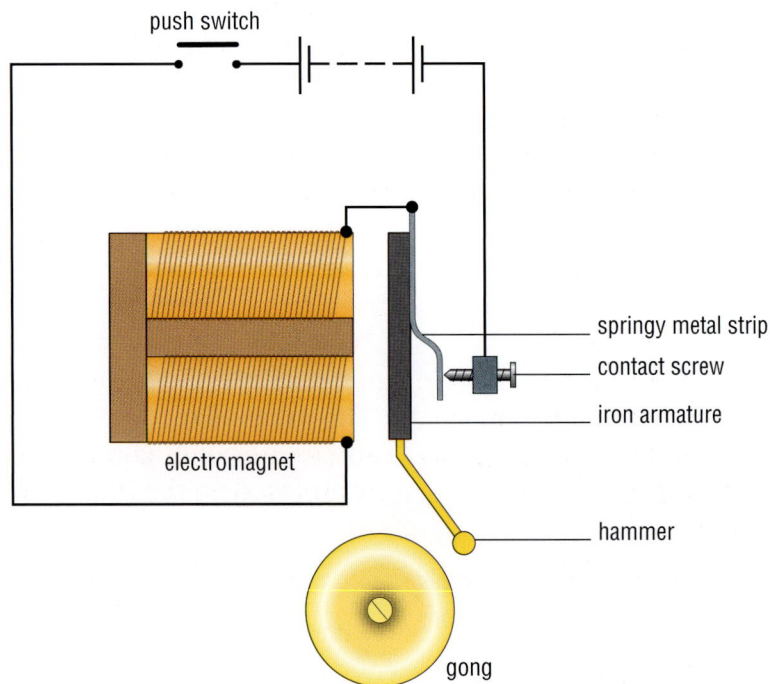

Figure 6.21 The circuit of an electric bell.

When the switch is pushed the current passes through the coil and the electromagnet pulls the armature to it. This makes the hammer strike the gong. When the armature is pulled to the electromagnet a gap develops between the springy metal strip and the contact screw and the circuit is broken. The current stops flowing and the electromagnet loses its magnetism. This makes the armature swing back to its original position. The springy metal strip and the contact screw now touch again and complete the circuit so the armature is pulled to the electromagnet once more. The bell is made to ring by the repeated beating of the hammer until the push switch is released.

18 Describe the changes that take place in the springy metal strip holding the armature when the current
a) flows, and
b) stops flowing.
Use the term strain force in your description.

◆ SUMMARY ◆

- Magnetic materials are attracted by a magnet; non-magnetic materials are not (*see page 54*).
- A magnet can attract or repel another magnet (*see page 55*).
- A bar magnet aligns itself in a north–south direction when it is free to move (*see page 55*).
- A magnet has a north-seeking pole and a south-seeking pole (*see page 55*).
- A magnet can be made by induction (*see page 56*).
- There are tiny regions called domains inside a magnet (*see page 56*).
- A magnetic field exists around a magnet (*see page 57*).
- The Earth has a magnetic field (*see page 58*).
- A wire with an electric current passing through it has a magnetic field around it (*see page 62*).
- An electromagnet is a magnet whose magnetism can be switched on and off by switching a current on and off (*see page 64*).
- A reed switch is opened and closed by a magnet (*see page 64*).
- An electromagnet is used to produce the repeated ringing of an electric bell (*see page 66*).

End of chapter questions

1 Why do magnets line up in a north–south direction?
2 Assess the importance of magnetism in the working of electrical devices in the home.

7 | *Light*

1 What is the luminous object which is providing light for you to read this book?

Light is a form of energy. It is a form of electromagnetic radiation (see page 30). Objects that emit light are said to be luminous while those that do not emit light are said to be non-luminous. Non-luminous objects can only be seen if they are reflecting light from a luminous source. The Moon is a non-luminous body – the 'moonlight' it produces is reflected sunlight.

Most luminous objects, such as the Sun, stars, fire and candle flames, release light together with a large amount of heat.

Figure 7.1   A bonfire is luminous: it radiates light and heat.

Light rays

Light leaves the surface of a luminous object in all directions but if some of the light is made to pass through a hole it can be seen to travel in straight lines. For example, when sunlight shines through a small gap in the clouds it forms broad sunbeams with straight edges (Figure 7.2). The path of the light can be seen because some of it is reflected from dust in the atmosphere. Similarly, sunlight shining through a gap in the curtains of a dark room produces a beam of light which can be seen when the light reflects from the dust in the air of the room.

Figure 7.2 Although the Sun radiates light in all directions, the sides of sunbeams seem almost parallel because the Sun is a very distant luminous object.

Smaller lines of light, called rays, can be made by shining a lamp through slits in a piece of card.

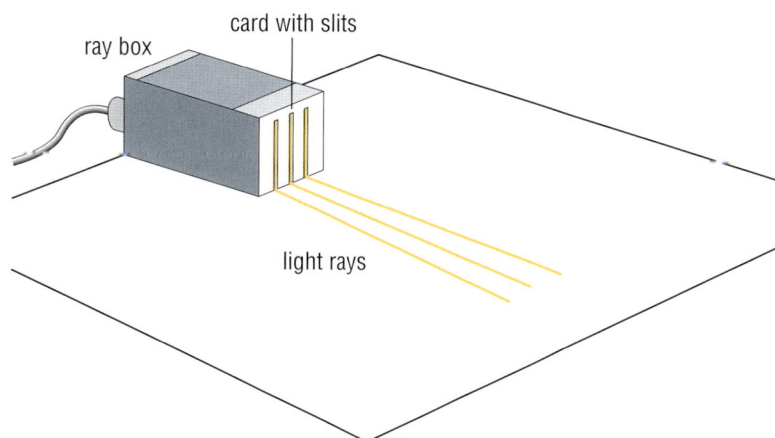

Figure 7.3 Making rays of light.

Classifying non-luminous objects

Non-luminous materials can be classified as transparent, translucent or opaque according to the way light behaves when it meets them. When light shines on a transparent material, such as glass in a window, it passes through it and so objects on the other side of it can be seen clearly.

When light shines on a translucent object, such as tracing paper, some of the light passes through but many light rays are scattered. Objects on the other side cannot be seen clearly unless they are very close to the translucent object.

When light shines on an opaque object none of the light passes through it.

Shadows

When a beam of light shines on an opaque object the light rays which reach the object are stopped while those rays which pass by the edges continue on their path. A region without light, called a shadow, forms behind the object. The shape of the shadow may not be identical to the shape of the object because the shadow's shape depends on the position of the light source and on where the shadow falls.

The size and intensity of the shadow depends on the size of the light source and the distance between the light source and the object. A small light source gives a sharp shadow that is equally dark all over. A larger light source gives a shadow with a dark central region and a lighter shadow surrounding it.

Shadows are also cast by the Moon and the Earth.

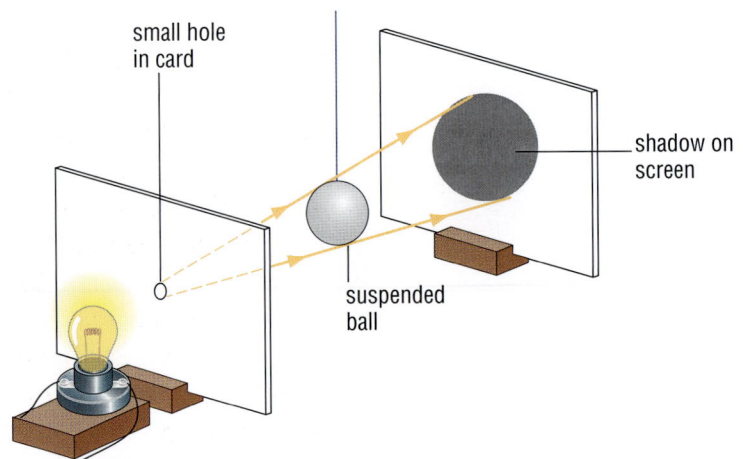

Figure 7.4   Formation of a shadow with a small light source.

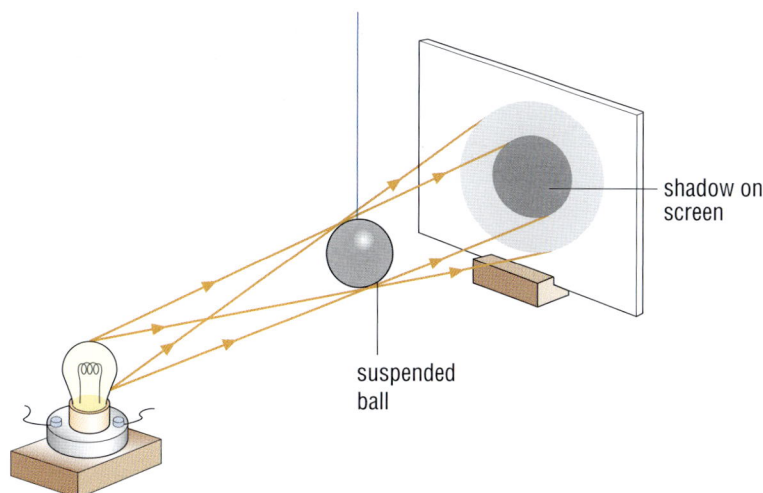

Figure 7.5 Formation of a shadow with a large light source.

For discussion

How might the shadow of a brick appear if light travelled in a curve from the light source?

If the light source is close to the object it makes a bigger shadow than if it is further away.

Reflecting light

Your bedroom is probably full of objects but if you were to wake in the middle of the night you could not see them clearly because they are not luminous. You can only see them by reflected light and unless your room is partially lit by street lights the objects will not be clearly seen until sunrise. The way light is reflected from a surface depends on whether the surface is smooth or rough.

Studying reflections

A few terms are used in the study of light which make it easier for scientists to describe their investigations and ideas. In the study of reflections the following terms are used:

- incident ray – a light ray that strikes a surface
- reflected ray – a light ray that is reflected from a surface
- normal – a line perpendicular (that is at 90°) to the surface where the incident ray strikes
- angle of incidence – the angle between the incident ray and the normal
- angle of reflection – the angle between the reflected ray and the normal
- plane mirror – a mirror with a flat surface
- image – the appearance of an object in a smooth, shiny surface. It is produced by light from the object being reflected by the surface.

The ways the incident ray, normal and reflected ray are represented diagrammatically is shown in Figure 7.6. The back surface of a mirror is usually shown as here, as a line with short lines at an angle to it.

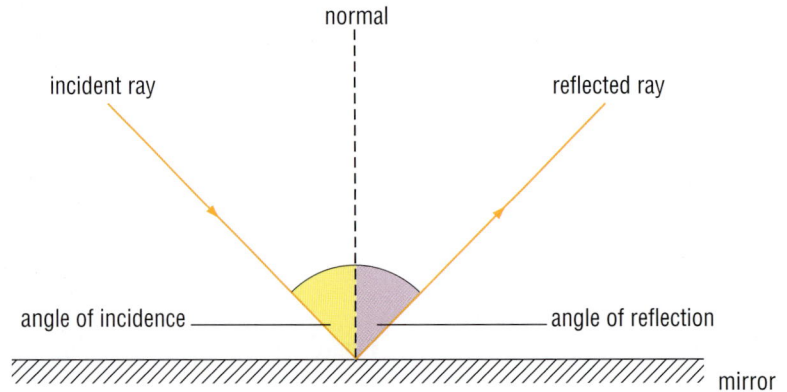

Figure 7.6   Reflection of light from a plane mirror.

The way light rays are reflected from a plane mirror can be investigated using the equipment shown in Figure 7.7.

2 Figure 7.8 shows three drawings made of the path of incident and reflected rays in an experiment using the apparatus in Figure 7.7. Use a protractor to measure the angles of incidence and angles of reflection. What do these drawings tell you about the process of reflection?

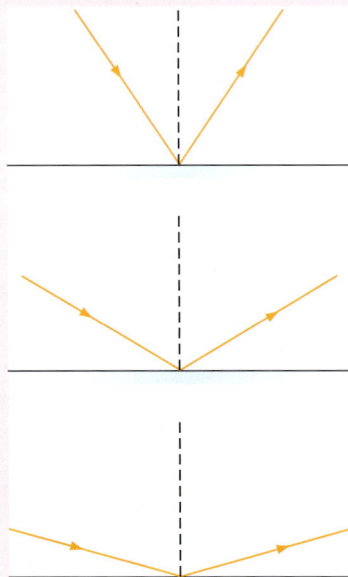

Figure 7.7   Investigating reflection from a plane mirror.

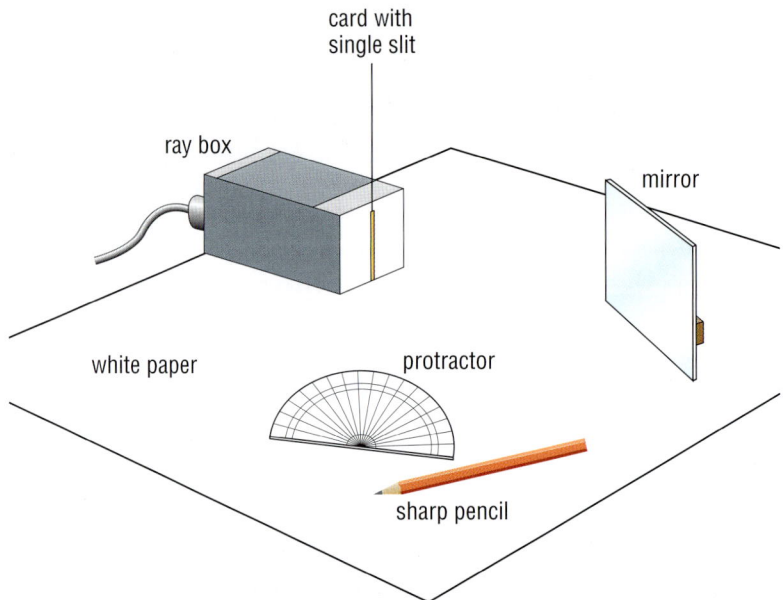

Figure 7.8

Objects with smooth surfaces

Glass, still water and polished metal have very smooth surfaces. Light rays striking their flat surfaces are reflected as shown in Figure 7.9. The angle of reflection is equal to the angle of incidence. When the reflected light reaches your eyes you see an image (Figure 7.10).

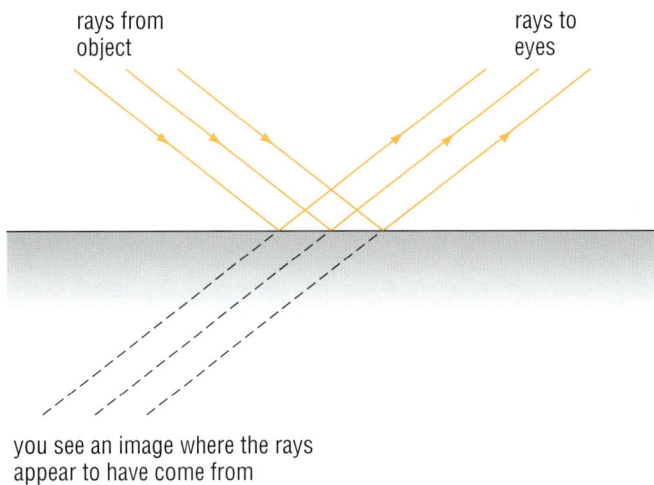

rays from object

rays to eyes

you see an image where the rays appear to have come from

Figure 7.9 Regular reflection from a smooth surface.

Figure 7.10 Light reflected from the smooth surface of a lake can produce an image in the water.

Two kinds of images

There are two kinds of images that can be formed with light. They are real images, such as those produced on a cinema screen by biconvex lenses and virtual images, which cannot be projected onto a surface but only appear to exist, such as those in a plane mirror or other smooth, shiny surface.

The virtual image of yourself that you see when you look in a plane mirror is the same way up as you are, is the same size as you are, and is at the same distance from the mirror's surface as you are but behind the mirror instead of in front of it. The main difference between you and your virtual image is that the virtual image is the 'wrong way round' – for example, your left shoulder appears to be the right shoulder of your virtual image.

Figure 7.11 Your image in a mirror is the wrong way round.

The periscope

Two plane mirrors may be used together to give a person at the back of a crowd a view of an event.

Figure 7.12 Some of the people in this scene are using periscopes to help them see over the crowd.

The arrangement of the mirrors in a periscope is shown in Figure 7.13.

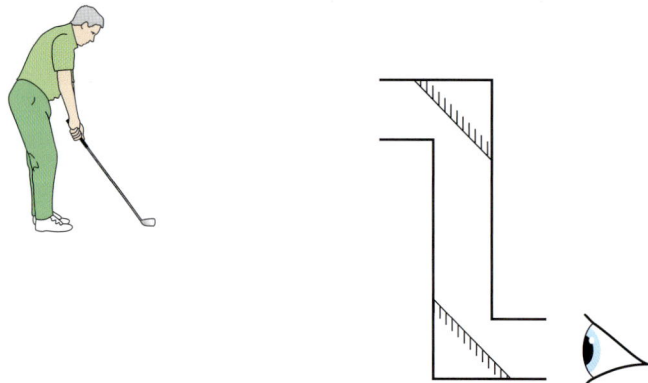

3 Copy Figure 7.13 and draw in the path of a ray of light travelling from the golfer to the eye.

4 Why is a periscope useful on a submarine?

Figure 7.13 A simple periscope.

Objects with rough surfaces

Most objects have rough surfaces. They may be very rough like the surface of a woollen pullover or they may be only slightly rough like the surface of paper. When light rays strike any of these surfaces the rays are scattered in different directions as Figure 7.14 shows.

You see a pullover or this page by the light scattered from its surface. You do not see your face in a piece of paper because the reflection of light is irregular, so cannot form an image.

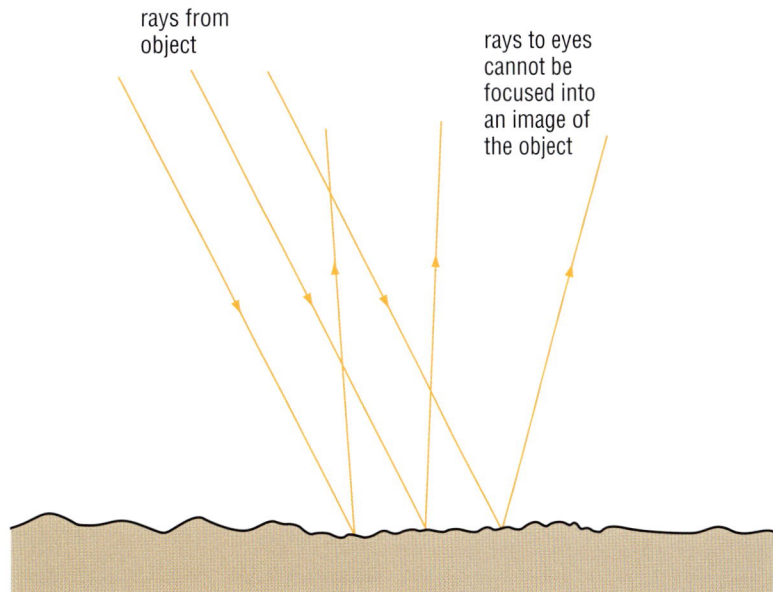

Figure 7.14 Light rays are scattered by a rough surface.

Passing light through transparent materials

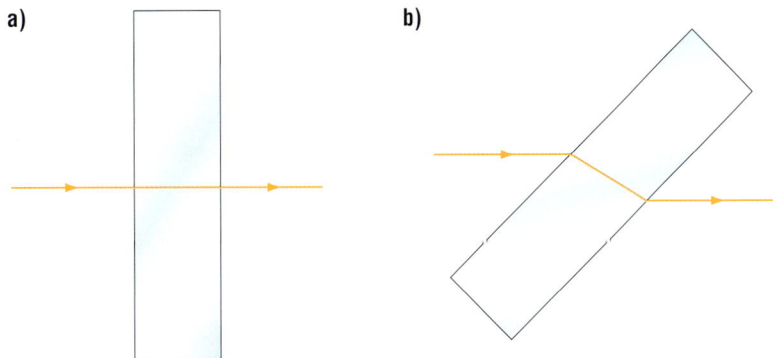

a)

b)

Figure 7.15 Light is refracted if the incident ray is not at 90° to the surface of the transparent material.

If a ray of light is shone on the side of a glass block as shown in Figure 7.15a the ray passes straight through, but if the block is tilted the ray of light follows the path shown in Figure 7.15b.

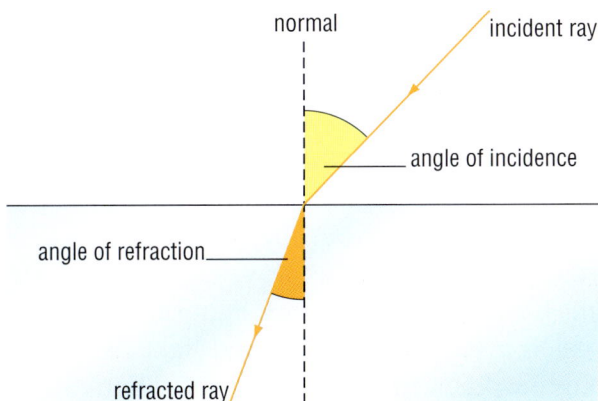

Figure 7.16 The angle of incidence and the angle of refraction.

This 'bending' of the light ray is called refraction. The angle that the refracted ray (see Figure 7.16) makes with the normal is called the angle of refraction.

The refraction of light as it passes from one transparent substance or 'medium' to another is due to the change in the speed of the light. Light travels at different speeds in different media. For example, it travels at almost 300 million metres per second in air but only 200 million metres per second in glass. If the light slows down when it moves from one medium to the other, the ray bends towards the normal. If the light speeds up as it passes from one medium to the next, the ray bends away from the normal.

Light speeds up as it leaves a water surface and enters the air. A light ray appears to have come from a different direction than that of the path it actually travelled (see Figure 7.17a and b). The refraction of the light rays makes the bottom of a swimming pool seem closer to the water surface than it really is. It also makes streams and rivers seem shallower than they really are and this fact must be considered by anyone thinking of wading across a seemingly shallow stretch of water. The refracted light from a straw in a glass of water makes the straw appear to be bent.

5 How is the reflection of a light ray from a plane mirror (see page 72) different from the refraction of a light ray as it enters a piece of glass?

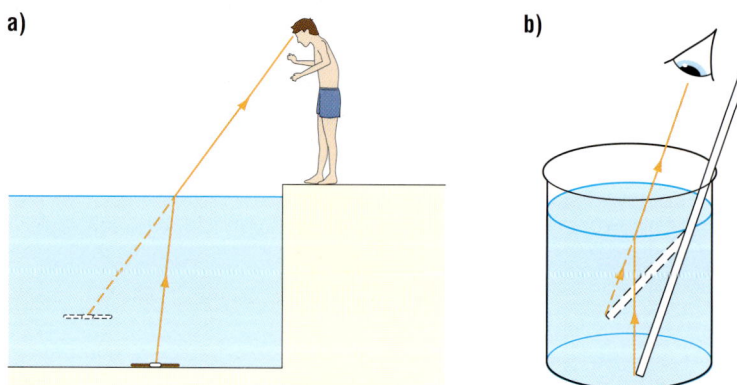

a)

b)

Figure 7.17 Refraction of light as it passes from water to air makes an object appear closer to the surface than it really is.

The prism

A triangular prism is a glass or plastic block with a triangular cross-section. When a ray of sunlight is shone through a prism at certain angles of incidence and its path is stopped by a white screen, a range of colours, called a spectrum, can be seen on the screen.

Light behaves as if it travelled as waves (see page 30). The 'white' light from the Sun contains light of different wavelengths which give different coloured light. When they pass through a prism the light waves of different

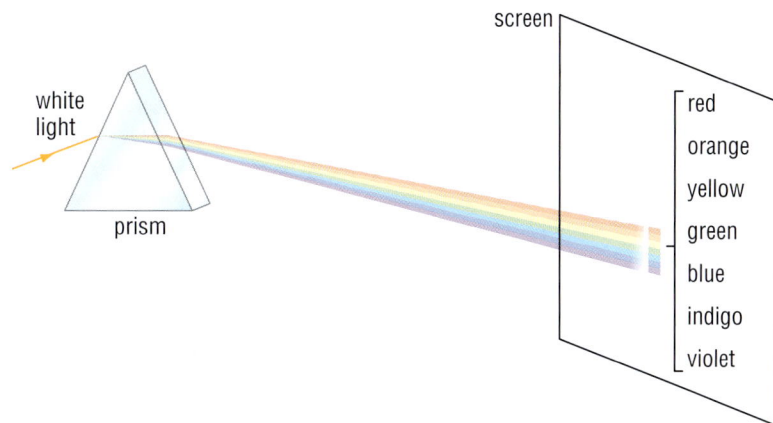

Figure 7.18 White light passing through a prism is split up into its constituent colours, forming a spectrum.

6 Look at Figure 7.18. Which colour of light has the shortest wavelength? Explain your answer.

wavelengths travel at slightly different speeds and are spread out, by a process called dispersion, to form the colours of the spectrum. The light waves with the shortest wavelengths are slowed down or refracted the most.

The rainbow

If you stand with your back to the Sun when it is raining or you look into a spray of water from a fountain or a hose you may see a rainbow. It is produced by the refraction and reflection of the Sun's light through the water drops. Figure 7.19 shows the path of a light ray and how the colours in it spread out to form the order of colours – the spectrum – seen in a rainbow.

Sometimes a second, weaker rainbow is seen above the first because two reflections occur in each droplet. In the second rainbow the order of colours is reversed.

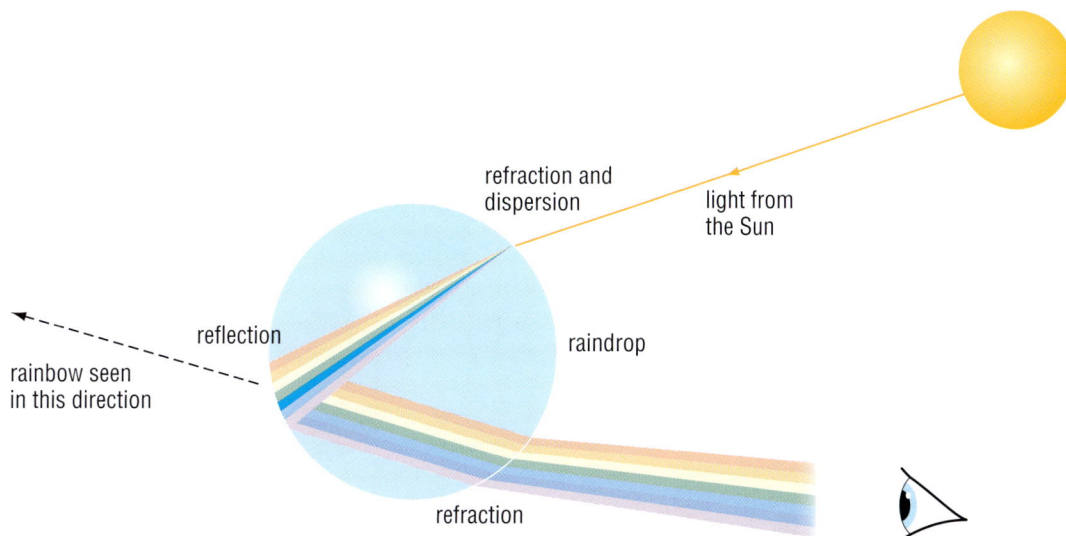

Figure 7.19 Formation of a rainbow.

Colour

Absorbing and reflecting colours

When a ray of sunlight strikes the surface of an object, all the different colours in it may be reflected or they may all be absorbed. If all the colours are reflected the object appears white; if all the colours are absorbed the object appears black.

Most objects, however, absorb some colours and reflect others. For example, healthy grass reflects mainly green and absorbs other colours.

Filtering colours

Sheets of coloured plastic or glass can filter the colours in light. They absorb some of the colours and allow other colours to pass through, producing different coloured light. For example, a blue filter allows only blue light to pass through.

Combining colours

When different coloured lights are combined it is found that all the colours can be made from different combinations of just three colours. They are red, green and blue, and are called the primary colours of light. These are different from the primary colours needed to make different coloured paint (see below).

When beams of the three primary colours are shone onto a white screen so that they overlap they produce three secondary colours of light and white light, as Figure 7.20 shows.

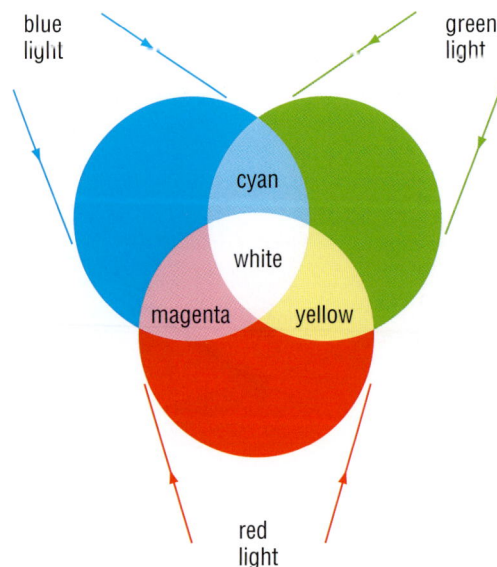

Figure 7.20 Overlapping beams of the primary colours form the secondary colours.

7 Name some everyday objects which
   a) reflect all the colours in sunlight,
   b) absorb all the colours in sunlight.

8 The colours on a television or computer screen are made by three different colours of substances called phosphors. They glow to release their colour of light. What do you think the colours of the phosphors are? Explain your answer.

9 Which primary colours overlap to produce
   a) yellow,
   b) magenta,
   c) cyan,
   d) white light?

For discussion

Identify the light source you are using for seeing things around you. Choose an object in the room. Describe the changes that take place in the light from when it leaves the source until it reaches your eyes from the object. Is it refracted through glass? Is it partially reflected from any surface? Which colours have been absorbed by the object?

Colours and paint

Three different colours of paint, ink or dye can be used to make almost all the other colours. These three colours are yellow, magenta and cyan. They are mixed together in different proportions to produce a wide range of colours, like those in the photographs in this book. Tiny dots of the three colours form the printed picture.

♦ SUMMARY ♦

♦ Light is a form of energy that is released from luminous objects (*see page 68*).

♦ Materials can be classified as opaque, translucent or transparent (*see page 70*).

♦ A shadow forms when light rays are stopped by an opaque object (*see page 70*).

♦ We see non-luminous objects by the light they reflect (*see page 71*).

♦ Light rays are reflected from a smooth surface at the same angle at which they strike it (*see page 72*).

♦ A real image can be formed on a screen but a virtual image cannot (*see page 73*).

♦ When light rays strike the surface of a transparent material at an angle to the perpendicular they are refracted (*see page 75*).

♦ A prism can split up sunlight into different colours of light (*see page 76*).

♦ The colour of an object we see depends on the colours of light that it absorbs and reflects (*see page 78*).

End of chapter questions

1 Describe what happens to light in a beam from the time it reaches the Earth from the Sun and shines upon a leaf, to when it enters your eye.

2 Imagine a kitchen. This is usually a room with many kinds of surfaces. There are also objects with many colours and some are transparent, translucent or opaque. Describe what happens to the sunlight as it strikes the different kinds of surfaces. Here are two examples to help you begin your answer. Think of four more and describe what happens to light when it strikes them too.
   • the white surface of a fridge
   • the black surface of an oven hob

8 Sound

You have probably performed some experiments on sound without knowing it. At some time most people have made a ruler vibrate by holding one end over the edge of a desk and 'twanging' it. The end of the ruler moves up and down rapidly and a low whirring sound is heard which becomes higher as you pull in the ruler from the edge of the desk.

Figure 8.1   Making a ruler vibrate.

From vibration to sound wave

Any object can make a sound wave when it vibrates. In practical work on sound you might use an elastic band, a guitar string or a tuning fork because they all vibrate easily. A vibration is a movement about a fixed point. This movement may be described as a to-and-fro movement or a backwards and forwards movement (Figure 8.2).

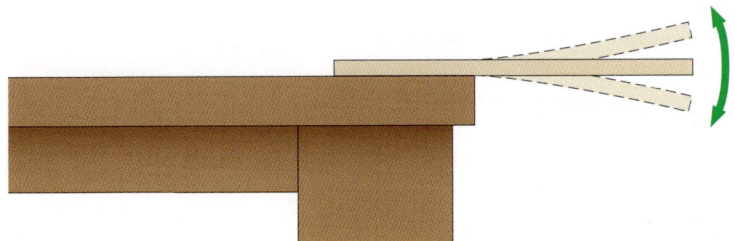

Figure 8.2   Vibration is a to-and-fro movement.

Figure 8.3 Producing sound by vibration.

Sound waves can travel in a gas, a liquid or a solid because they all contain particles (see page 45). When an object vibrates it makes the particles next to it in the gas, liquid or solid vibrate too. For example, when an object vibrates in air it pushes on the air particles around it.

As the vibrating object moves towards the air particles it squashes them together. The particles themselves are not compressed but the pressure in the air at that place rises because the particles are closer together (Figure 8.4a).

As the object moves away from the air particles next to it, it gives them more space and they spread out and the pressure at that place falls (Figure 8.4b).

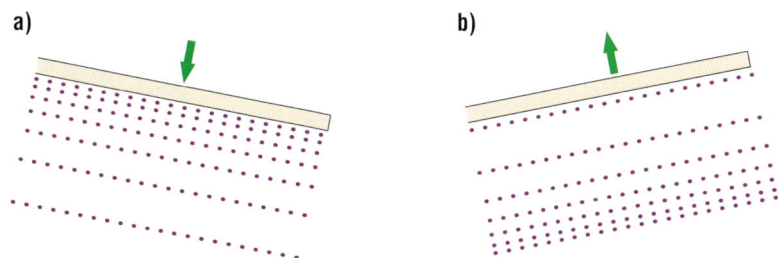

Figure 8.4 A vibrating object causes pressure variations in the air around it.

As the object vibrates the air particles nearby also move backwards and forwards and they in turn cause other air particles further away to squash together and then spread out. This makes alternate regions of high and low pressure which travel through the air away from the vibrating object (Figure 8.5).

1 When a table tennis ball on a thread is made to touch a vibrating prong of a tuning fork the ball swings backwards and forwards. How can this demonstration be used to explain how sound waves are made?

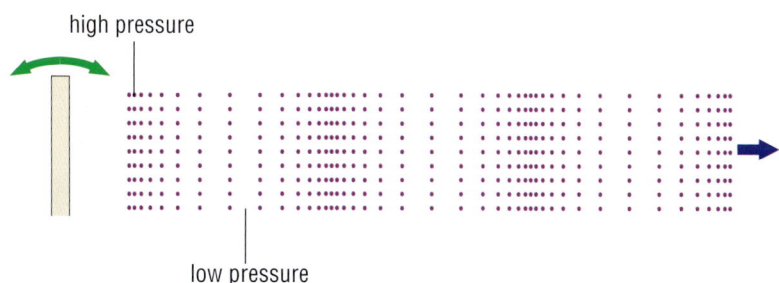

high pressure

low pressure

Figure 8.5 Regions of high and low pressure move away from the vibrating object.

If these changes in pressure were plotted on a graph they would make a waveform similar to that shown in Figure 8.9 (page 83). The waves of sound move out from the vibrating object in all directions.

Sound waves are generated and travel in liquids and solids in the same way as they do in gases. The particles in liquids and solids are held close together by forces of attraction (see page 45). In a liquid however the particles are further apart than in a solid and can move over one another. Sound travels very well through a liquid. It moves faster and further than it does in a gas.

Figure 8.6 These whales communicate by sound waves.

The humpback whale emits a series of sounds called songs which travel thousands of kilometres through the ocean. It uses its songs to communicate with other whales.

When sound travels through a solid it moves even faster than through a liquid because of the close interactions of the particles. However, the sound does not travel so far.

Figure 8.7 This snake is listening for vibrations in the ground.

A snake detects vibrations in the ground with its lower jaw bone. The bone transmits the vibrations to the snake's ears and helps the snake listen for the footsteps of its prey.

Sound waves cannot pass through a vacuum because it does not contain any particles. Figure 8.8 shows an experiment that demonstrates this. As air is drawn out of the bell jar with a pump, the sound of the bell becomes quieter. When a vacuum is established in the bell jar the bell cannot be heard although the hammer can be seen striking it.

2 Why is it that a bell in a sealed bell jar
  a) can be heard when the jar is full of air, but
  b) cannot be heard when a vacuum is created in the jar?

Figure 8.8 Sound cannot be heard through a vacuum.

Describing the wave

Figure 8.9 shows the different positions particles can occupy when a sound wave is produced. This type of graph is called a displacement/distance graph.

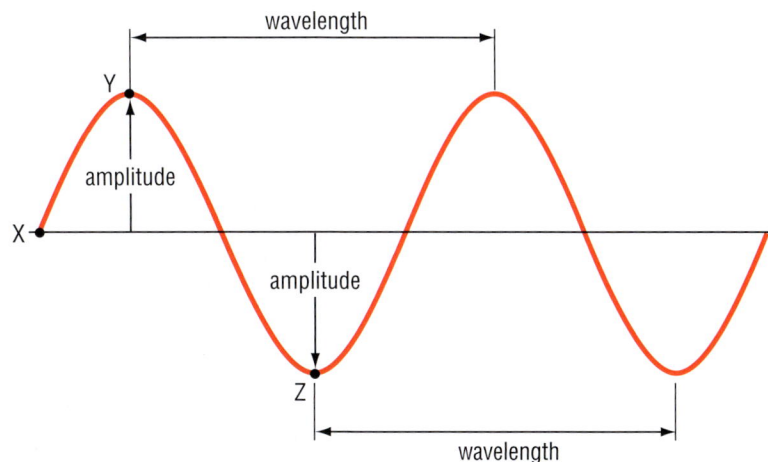

Figure 8.9 Displacement/distance waveform for a sound wave.

A particle at position X (Figure 8.9) is moving through the 'rest' position, a particle at Y has moved the maximum distance in one direction and one at Z has moved the maximum distance in the other direction.

Two characteristics of the wave that can be seen in Figure 8.9 are the amplitude and wavelength. The amplitude is the height of the crest or the depth of the trough and shows the maximum displacement of the particles from their rest position. The wavelength is the distance from the top of one crest to the top of the next crest, or from the bottom of one trough to the bottom of the next trough.

3 Can you think of other ways of describing the wavelength of a wave?

Detecting sound waves

The ear is the organ of the body that detects sound waves. It is divided into three parts – the outer ear, middle ear and inner ear (Figure 8.10).

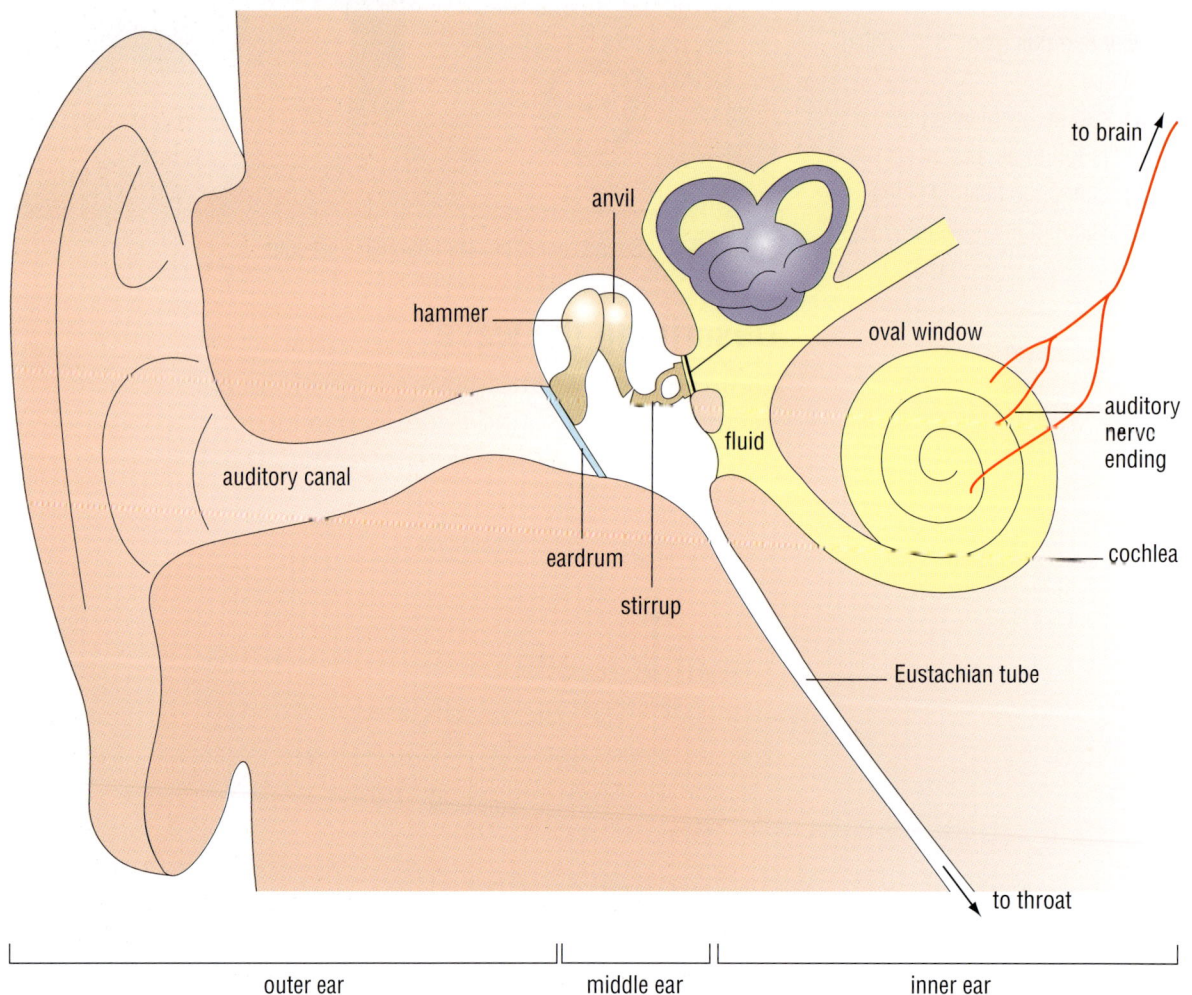

Figure 8.10 Structure of the ear.

The outer ear

When sound waves reach the outer ear some pass directly down the middle of the tube called the auditory canal. Some waves which strike the outer part of the ear are reflected into the auditory canal. At the end of the auditory canal is a thin membrane which stretches across it. This is called the eardrum. When sound waves reach the eardrum they push and pull on it and make it vibrate.

Some predatory animals, such as cats, can turn their outer ears forwards to detect sounds from prey in front of them. Some prey animals, such as rabbits, can turn their outer ears in many directions about their head to listen for approaching predators.

4 Why do people put a hand to their ear when they are listening to someone who is whispering?

Figure 8.11   This rabbit is able to turn its outer ears to capture sound waves from all directions.

The middle ear

In the cavity of the middle ear are three bones. They are called the hammer, anvil and stirrup, after their shapes. The ear bones form a system of levers. When the eardrum vibrates its movements are amplified by the lever system. The oval window on which the stirrup bone vibrates has a much smaller area than that of the eardrum. This difference in area between the eardrum and the oval window causes the vibrations of the eardrum to be amplified as they enter the inner ear and set up vibrations in the fluid there.

5 Why do people go partially deaf when they have a very heavy cold and their Eustachian tubes become blocked?

The middle ear also has a tube, the Eustachian tube, which connects to the throat. When we swallow the tube opens and the air in the middle ear is connected to air outside the body. This brief connection allows the air pressure in the ear to adjust to the air pressure outside the body (see page 151). This balancing of the air pressure allows the eardrum to vibrate as freely as possible.

The inner ear

The inner ear is filled with a fluid. The vibrations of the stirrup set up waves in the fluid. There is a membrane with delicate fibres in the cochlea. Each fibre only vibrates in response to a sound wave with a particular pitch (see pages 88–89). When a fibre vibrates it stimulates a nerve ending and a nerve impulse or message is sent to the brain where we become aware of the sound.

The cathode ray oscilloscope

The cathode ray oscilloscope or CRO can be used to investigate sound waves. A CRO contains a cathode ray tube. At one end of the tube is a screen. The cathode rays make a spot on the screen which can be made to move from left to right. When a microphone is attached to the CRO, sound waves can be displayed on the screen. Sound waves striking the microphone set up electrical signals in it which pass to the CRO. When the CRO receives them, the spot moves up and down as it travels across the screen and makes a waveform that can easily be seen.

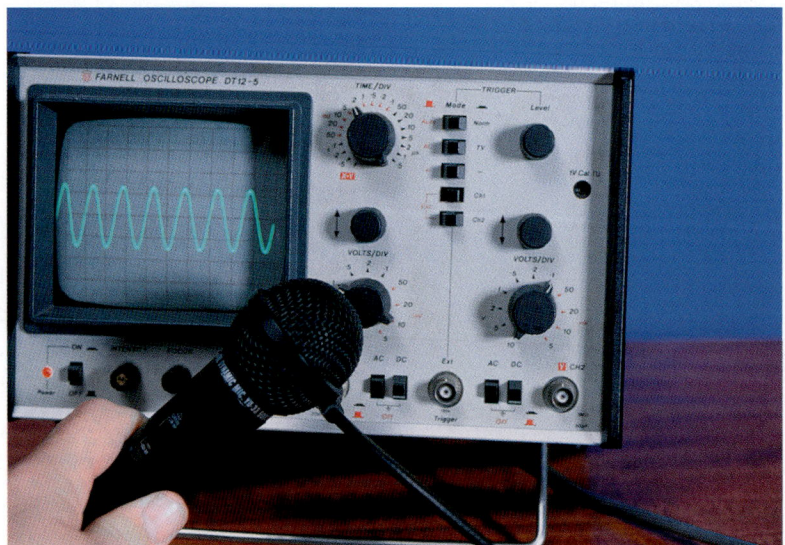

Figure 8.12 A cathode ray oscilloscope displaying a sound wave.

The loudness of sounds

The loudness of a sound is related to the movement of the vibrating object. If an object only moves a short distance to and fro from its rest position, it will produce a sound wave with only a small amplitude and the sound that is heard will be a quiet one.

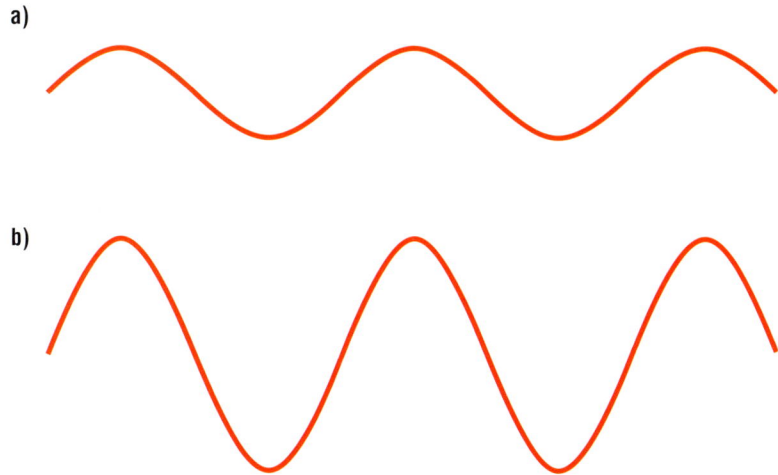

a)

b)

Figure 8.13 Displacement/distance waveform of **a)** a quiet sound and **b)** a loud sound.

If an object moves a large distance to and fro from its rest position, it will produce sound waves with a large amplitude and the sound that is heard will be a loud one. The loudness of sounds is measured in decibels (see Table 8.1).

Table 8.1 Loudness values of sounds.

Sound	Loudness/decibels
The sound hurts	140
A jet aircraft taking off	130
A road drill	110
A jet plane overhead	100
A noisy factory floor	90
A vacuum cleaner	80
A busy street	70
A busy department store	60
Normal speech	55
Voices in a town at night	40
A whisper	20
Rustling leaves	10
Limit of normal hearing	0

Figure 8.14 Wearing ear protection in a noisy boiler room prevents ear damage.

Loudness and energy

Sound energy passes through the air as the particles move to and fro. When a wave with a small amplitude is generated, a small amount of energy passes through the air. When a wave with a large amplitude is generated a large amount of energy passes. The energy of a sound wave is converted into other forms such as movement energy in the eardrum and ear bones.

Loudness and deafness

The vibrating air particles of a very loud sound can produce such a strong pushing and pulling force on the eardrum that a hole is torn in it. The eardrum is said to be perforated. It no longer vibrates efficiently and the person loses his or her hearing. The eardrum can heal and normal hearing can be restored.

If a person is exposed to a very loud sound or a particular note for a long period of time he or she will no longer be able to hear it. This is due to permanent damage to a nerve ending in the cochlea. People who perform in pop groups are at risk of developing this kind of deafness, called nerve deafness. In time they may be unable to hear a range of notes which they frequently used in their music. People who work in noisy surroundings, such as airport workers or metal workers in a factory, wear ear protection in the form of ear muffs which cover the ears and reduce the amount of sound energy entering the ears.

A common form of partial deafness, which is not related to the loudness of a sound, is the development of ear wax in the outer ear. This stops sound waves reaching the eardrum. The wax can be removed with warm water under the medical supervision of a nurse.

Some people have growths of tissues in their middle ears which stop the ear bones moving freely. They may be prescribed with a hearing aid. This contains a microphone and amplifier and compensates for some of the loss of amplification that was provided by the ear bones.

The pitch of a sound

You probably have an idea about the pitch of a sound even if you don't know the word. You might describe a sound as a high or a low sound, which really means a

6 What kind of ear damage might be caused by a loud explosion? Explain your answer.

For discussion

To prevent ear damage, how should you use earphones on a CD player? Where should you dance at a disco, and where should you sit or stand at a pop concert?

How far do you follow the advice you have given in answer to the above?

high-pitched or a low-pitched sound. For example, when you say 'bing' you are making a higher-pitched sound than when you say 'bong'.

The pitch of the sound an object makes depends on the number of sound waves it produces in a second as it vibrates. This number of waves per second is called the frequency. The frequency of a sound is measured in hertz (abbreviation Hz). The higher the frequency of the wave, the higher the pitch of the sound.

The graphs in Figure 8.15 show the positions that particles occupy at different times as the wave passes. These graphs are called displacement/time graphs. The higher frequency waves have a shorter wavelength than the lower frequency waves. Sound waves share this property with light waves (see page 30).

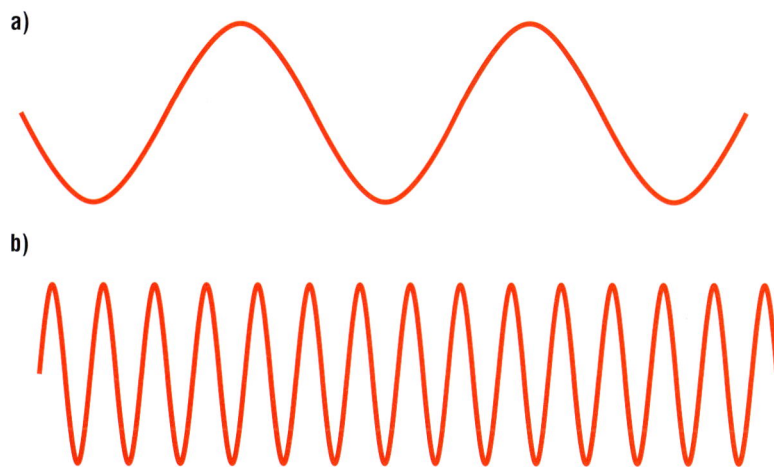

a)

b)

Figure 8.15 Displacement/time waveform of
a) a low frequency sound and **b)** a high frequency sound.

7 The following are three frequencies of sound waves: 1800 Hz, 50 Hz, 10 000 Hz
  a) Which has the highest pitch and which has the lowest pitch?
  b) What does Hz stand for?

The ear of a young person is sensitive to frequencies in the range 20 to 20 000 hertz, but the ability to detect the higher frequencies decreases with age. Some people may have a restricted range of hearing due to nerve damage. They may not be able to hear some low-pitched or high-pitched sounds.

Experiments on sound

In the past many scientists have performed experiments to find the speed of sound. Isaac Newton (1642–1727) investigated the speed of sound by measuring the time between a sound being made and its echo from a wall being heard. Other scientists measured the time taken between seeing a distant cannon fire and hearing its sound.

Figure A Measuring the speed of sound in air.

The speed of sound in water was investigated using the apparatus shown in Figure B. The experiment was performed at night. When the lever was pulled down both the arm carrying the bell hammer and the device carrying the match moved.

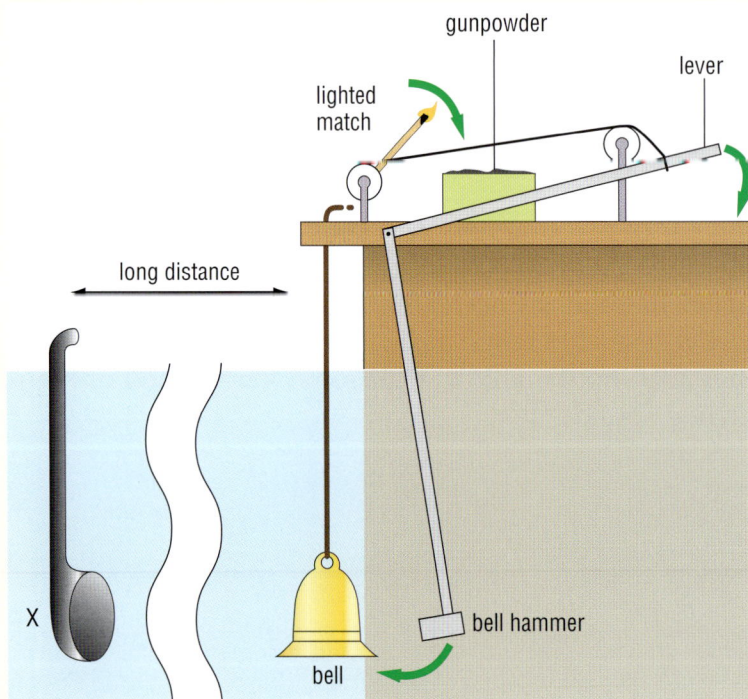

Figure B Measuring the speed of sound in water.

1 What is an echo?
2 What measurement besides time needs to be taken in all the experiments to determine the speed of sound?
3 What is the purpose of
   a) the gunpowder, and
   b) the apparatus marked X in Figure B?
4 Why do you think the experiment to find the speed of sound in water was done at night?

◆ SUMMARY ◆

♦ Sounds are made by vibrating objects (*see page 80*).

♦ Sound travels through materials as waves of vibrating particles (*see page 81*).

♦ There are three parts to the ear. Each part plays an important role in hearing (*see page 84*).

♦ A cathode ray oscilloscope is used to investigate sound waves (*see page 86*).

♦ The loudness of a sound is related to the amplitude of its waves (*see page 87*).

♦ The pitch of a sound is related to the frequency of its waves (*see page 88*).

End of chapter question

1 Describe how the vibration of a ruler is detected in the inner part of your ear.

9 Energy transformations

You are living at this moment because of energy transformations taking place in your body. Stored chemical energy in your food is transformed to kinetic (movement) energy when you raise your hand to turn the pages of this book. Some of the stored chemical energy is also transformed into heat to keep your body warm.

Figure 9.1 Energy transformations in everyday life.

All the changes you can detect around you are due to energy transformations. If you are reading this book by an electric light, electrical energy is being transformed into light energy so that you can see the words. If you can hear someone shuffling about in their seat next to you, some of the kinetic energy of their body is being transformed into sound energy that reaches your ears.

On page 34 energy transformations were described more simply as energy changes. If you look back at those pages you will see that at each energy change or transformation some energy is always lost as heat (thermal) energy.

Energy transformations are vital to survival. From the earliest times people have needed enough stored chemical energy in their bodies to change to kinetic energy to move around and find food. If they could not find enough food (and store it as chemical energy for later use), they simply starved to death.

In time people came to live together in groups and developed machines and other devices to make life easier. Figure 9.2 shows a city about 100 years ago.

1 If you are wearing a watch, what energy changes are taking place in it right now?

A city in the past

Figure 9.2 A city scene from about 100 years ago.

The insides of the buildings were lit by candles, oil lamps and gas lights. Coal and wood were used as fuel on fires to keep the buildings warm in winter. The streets were lit by gas lights. People and goods were transported through the streets by carriages and carts pulled by horses and donkeys. Many people arrived or left the city by trains pulled by steam locomotives. Some people walked.

A city today

Figure 9.3 A modern city.

Figure 9.3 shows a city today. The insides of the buildings are lit by electricity. Some buildings have air conditioning powered by electricity to keep them cool in summer. They also have heaters powered by electricity to keep them warm in winter. Cars, vans, trucks and buses are used to transport people and goods. These vehicles are powered by petrol or diesel engines and are also used by people travelling to and from the city. Many people may also use trains pulled by a locomotive with an electric motor. Some people may visit the city by aeroplanes, which use a fuel called kerosene which is similar to petrol.

2 How have the sources of energy for light and movement in cities changed over the last hundred years?

Measuring work

On page 29 you can see how the words energy and work are linked together in science.

Work occurs when a force moves an object through a distance. The amount of work done can be calculated using this equation.

$$\text{work} = \text{force} \times \text{distance}$$

The strength of a force is measured in newtons (N) and the distance is measured in metres (m).

Work, then, is measured in newton-metres (Nm). This unit of work is called the joule (J). Energy is also measured in joules.

Making a calculation

The weight of an object is measured in newtons (see page 115). The weight is the pull of the Earth's gravity on the object. When you lift an object you are increasing its potential energy (see page 33). The work done to do this is found by measuring the height through which you lift the object multiplied by the force that acts on it due to gravity (its weight).

For example, if you raised a bag of sand that weighed 10 N 1 metre off the ground, you would need to expend 10 N × 1 m = 10 joules of energy to do it, or do 10 joules of work.

Energy transfer diagrams

Energy transformations can be shown by energy transfer diagrams. There are three parts to an energy transfer diagram: (i) an arrow showing the energy input, (ii) a box showing an energy converter or transducer and (iii) arrows showing energy output. Here are some examples of energy transfer diagrams.

1 Releasing a catapult

strain energy → | catapult | → kinetic energy

2 Burning gas in a Bunsen burner

stored chemical energy → | Bunsen burner | → heat energy / → light energy

3 Blowing up a balloon

kinetic energy → | balloon | → strain energy

4 Taking a photograph

light energy → | camera | → chemical energy

3 How much energy would you use to lift a 10 N sand bag
 a) 2 m,
 b) 2.5 m,
 c) 3 m above the ground?
4 A girl weighs 400 N and climbs a staircase 9 m high. How much energy will she have used when she reaches the top?

5 Draw energy transfer diagrams for
 a) winding up a clockwork car,
 b) letting a clockwork car run,
 c) letting a battery-powered car run.
6 Diagram 4 shows the energy transfer for a camera using a film. What would the energy transfer diagram be for taking a picture with a digital camera in a mobile phone?
7 Diagrams 1, 3 and 4 show the main energy changes. However, there should be a second energy output for complete accuracy. What is this output? Explain your answer. (You may like to turn to page 34 to help you answer.)

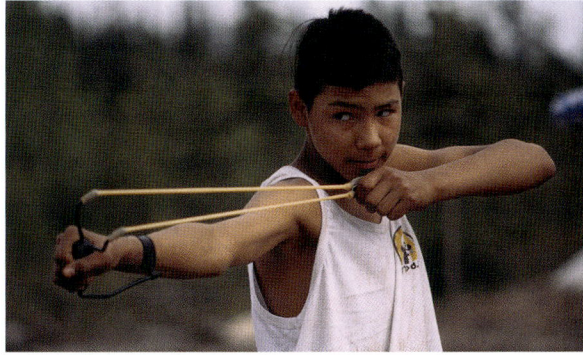
Figure 9.4 Releasing a catapult.

Figure 9.5 Blowing up a balloon.

Sankey diagrams

A Sankey diagram is a second kind of diagram that shows energy changes. It features arrows of different widths. The width of the arrow indicates the amount of energy it represents.

Energy in a car engine

A car engine has an energy input of 200 kJ. The energy output is 80 kJ of kinetic energy and 120 kJ of heat energy.

8 Draw a Sankey diagram for a light bulb that has an energy input of 100 J and an energy output of 10 J of light energy and 90 J of heat energy.

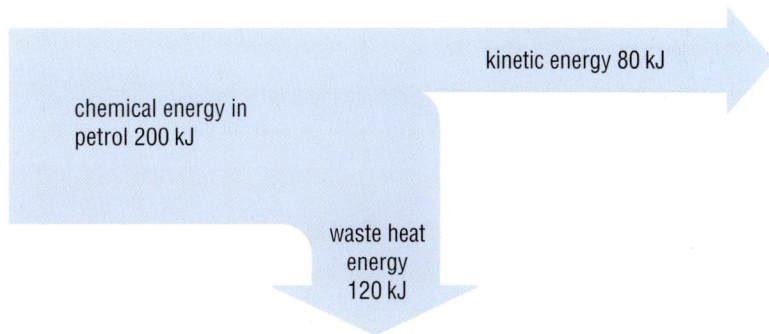

Figure 9.6 This Sankey diagram shows what happens to the energy passing through a car engine every second.

Energy from the Sun

Figure 9.7 on page 96 is a diagram showing the path of energy reaching the Earth from the Sun.

Plants and energy

When a seed germinates its skin breaks open and the tip of the root pops out. Energy stored inside the seed is used as the root grows and seeks out water. Stored energy is also used by the growing shoot.

The shoot grows up through the soil and eventually reaches the surface. Energy from below, in the seed, is used as the shoot sends out leaves.

173 × 10^{12} kJ
of energy arrive
every second

30% of the energy
is reflected back
into space

47% of the energy
is absorbed by
the atmosphere

23% of the energy makes
the water circulate in the
water cycle

less than 1% of the energy is used
to produce winds and currents

0.02% of the energy is used
by plants in photosynthesis

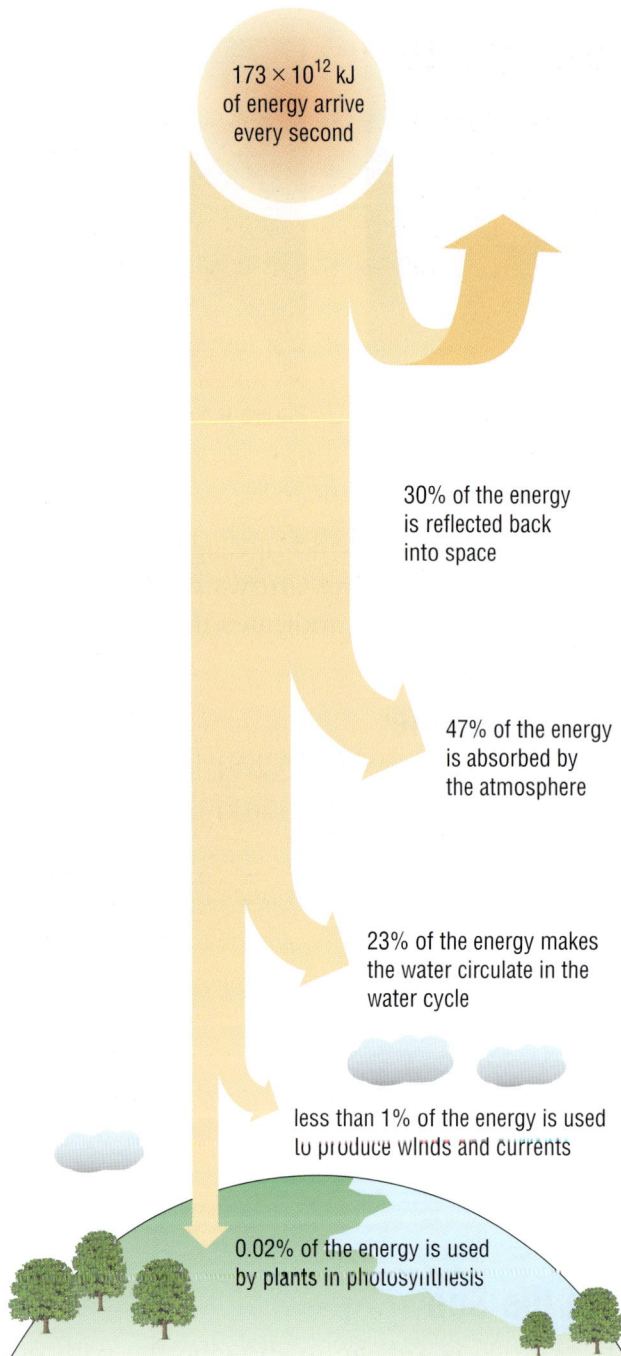

Figure 9.7 Sankey diagram of energy from the Sun.

Some of the light energy falling on the leaves is trapped inside them. It is converted into stored chemical energy as the plant makes food using water from the soil and carbon dioxide from the air. This process of making food using light energy is called photosynthesis. You can read more about it in Chapter 11 of *Checkpoint Biology*.

The chemical energy stored in a plant is transferred to a herbivorous animal when it eats it. The herbivorous animal then has a store of chemical energy which it uses to keep itself alive and to move about. Carnivorous animals feed on herbivorous animals. They take in stored chemical energy when they eat their prey. Thus the energy stored and used by animals originally came from light energy from the Sun that was trapped by plants.

9 Draw an energy transfer diagram for photosynthesis.

10 What do you think might happen to a seedling shoot if it was growing from a seed that had been planted too deeply?

Figure 9.8 Germinating seed.

Figure 9.9 Seedling during growth.

Energy and ourselves

Figure 9.10 Food cooking on a barbecue.

Have you ever cooked meat on a barbecue? If you have, you might have seen fat dropping through the grill and bursting into flames as it hit the hot charcoal below. If you did not take enough care when cooking the meat you may even have seen it catch fire too. The meat and fat burn because they contain chemical energy just like the charcoal fuel that is heating them. In fact, all foods contain energy and you can find out more about energy in foods on page 99 of *Checkpoint Biology* in this series.

If you look at a packet of food, you will find an information box. It tells you about the ingredients used and the nutrients present. It also tells you about the amount of energy in a 100 g mass of the food. The units used to measure energy in food are the joule and kilojoule. They are the same units as those used to measure energy and work in scientific investigations.

The chemical energy in food is released in a process called respiration. Oxygen is needed for this process and your body takes it in from the air you breathe into your lungs. When the energy is released, carbon dioxide and water are produced. The carbon dioxide is released into the air when you breathe out. The water is used in your body or released in sweat and urine. Most of the energy that is released in your body is used for movement and for keeping the body warm.

Nutrition

Typical Composition	This pack (450g) provides	100g (3¹/₂oz) provide
Energy	2610kJ	580kJ
	621kcal	138kcal
Protein	13.2g	2.9g
Carbohydrate	82.3g	18.3g
of which sugars	18.0g	4.0g
Fat	26.6g	5.9g
of which saturates	13.5g	3.0g
mono-unsaturates	10.4g	2.3g
polyunsaturates	2.7g	0.6g
Fibre	7.2g	1.6g
Sodium	1.8g	0.4g

A serving (450g) contains the equivalent of approx. 4.5g of salt.

Figure 9.11 Food packet label.

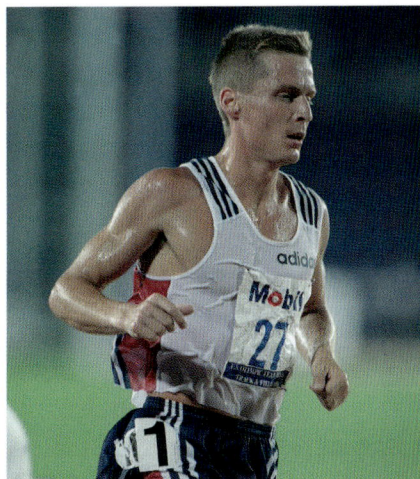

11 Draw an energy transfer diagram for the body. The energy input is the stored chemical energy in food.

Figure 9.12 This athlete is releasing a lot of energy.

Generating electricity

Electrical energy is a very useful form of energy because it is easy to generate and can be transported quickly to wherever it is needed. These properties make it the most widely used form of energy in the highly developed countries of the world.

Michael Faraday (1791–1867), an English physicist, discovered that an electric current could be made to flow in a wire if the wire was made to move through a magnetic field (see page 57). This principle is used to generate electricity in a bicycle dynamo and in a power station generator.

The bicycle dynamo

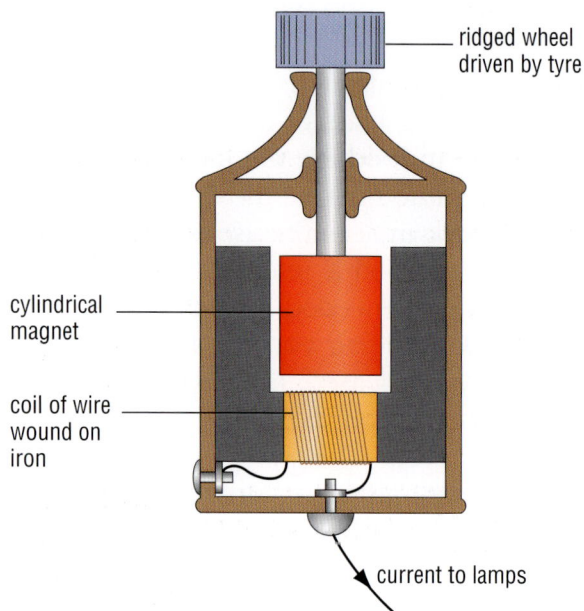

ridged wheel driven by tyre

cylindrical magnet

coil of wire wound on iron

current to lamps

Figure 9.13 Inside a dynamo.

12 What would be the problem if dynamos were the only way electricity was provided in bicycle lamps?

A bicycle dynamo is an electrical device which is clamped onto the frame of a bicycle close to a tyre. It has a wheel on top which can be made to touch the tyre. The inside of a dynamo is shown in Figure 9.13.

When the dynamo wheel is in contact with the tyre it rotates as the bicycle wheel turns. Inside the dynamo the magnet turns and its field sweeps through the wires, generating an electric current which lights the bicycle's lamps.

The power station generator

Inside a power station there is a generator consisting of a huge electromagnet surrounded by coils of wire. The electromagnet is attached to a shaft to which turbine blades are attached (Figure 9.14). When the turbines are made to spin, the electromagnet also spins, generating a current of electricity in the surrounding coils of wire.

In about two-thirds of the world's power stations water is heated to make steam. This takes place in a boiler. The energy that the water molecules receive increases their kinetic energy so much that they move apart from each other to form a gas – steam. The steam expands rapidly and exerts a force which drives it from the boiler to the turbine blades. Here as much as possible of the steam's kinetic energy is passed to the

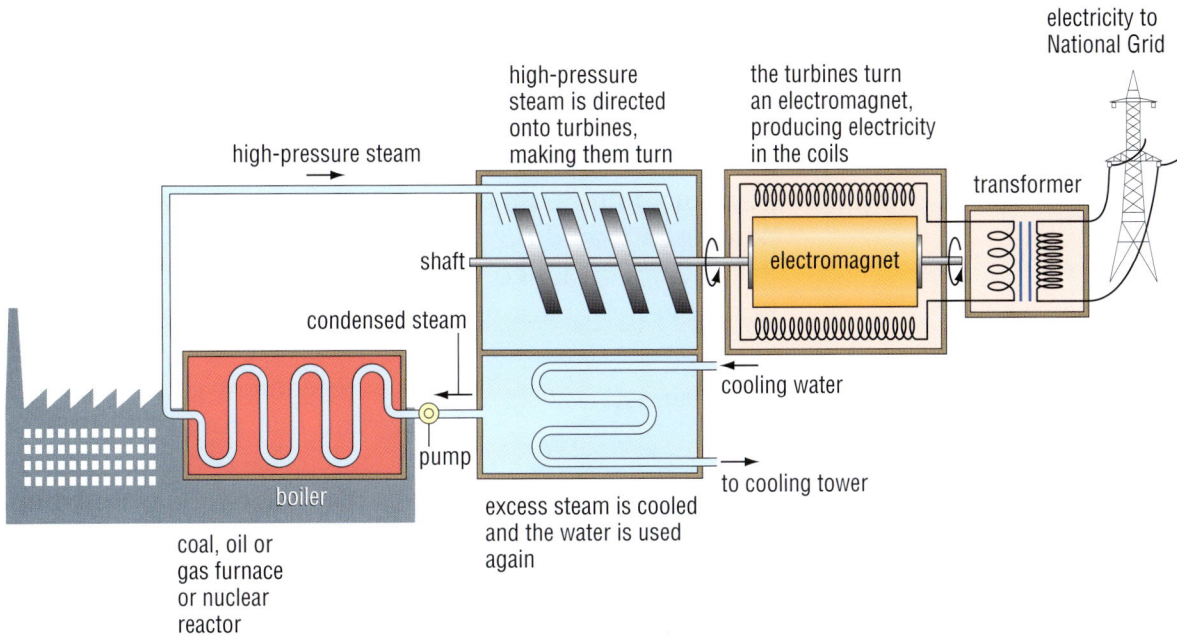

electricity to
National Grid

high-pressure
steam is directed
onto turbines,
making them turn

the turbines turn
an electromagnet,
producing electricity
in the coils

high-pressure steam

transformer

shaft

electromagnet

condensed steam

cooling water

pump

to cooling tower

boiler

excess steam is cooled
and the water is used
again

coal, oil or
gas furnace
or nuclear
reactor

Figure 9.14 *The parts of a power station.*

turbine blades as the steam pushes past them, making
the blades spin on the central shaft.

The generator's electromagnet is connected to the end
of the shaft. As it spins using kinetic energy from the
turbine blades it generates a current of electricity in the
coils of wires surrounding it. The electricity flows away
from the power station to towns and cities in overhead
power lines or underground cables.

13 Draw an energy transfer
diagram
  a) for the power station boiler,
  b) for the spinning
  electromagnet.
14 Describe how life in your home
would change if none of the
electrical appliances worked.

For discussion
How would your daily life have
to change if electricity was no
longer generated at any power
stations in your country?

Figure 9.15 Turbine assembly.

Conservation of energy

Imagine that you have a piece of string with a small weight on the end. It can be used as a pendulum. If you held it up in front of you and pulled the weight towards your face until it almost touched your nose the pendulum would be ready to swing. You might think that when you let go of the weight it would swing away from you then back again and perhaps hit your nose. If you were to try this experiment you would find that your nose was safe. The pendulum would not reach your nose when it swung back. In fact as the weight swung to and fro it would approach your nose less and less until it stopped.

It would seem that as the pendulum swung back the first time some energy was lost. It is true that pendulums do lose energy as they swing but the energy is not destroyed. As the string and weight move through the air, they rub against the air particles. This rubbing causes heat energy to be released from them, just like the heat energy that is released from your hands when you rub them together. With each swing of the pendulum, more heat energy is lost until the energy has all left the pendulum and it hangs vertically, motionless.

This experiment shows that energy is not destroyed, it just changes form. Just as energy is not destroyed so it is not created either. The energy used in the body or a machine simply comes from somewhere else in the Universe. This discovery about energy led to the Law of Conservation of Energy which says that energy cannot be made or destroyed, it can only be changed from one form to another.

Some inventors have tried to defy the Law of Conservation of Energy by designing and even making perpetual motion machines. Figure 9.17 shows an example.

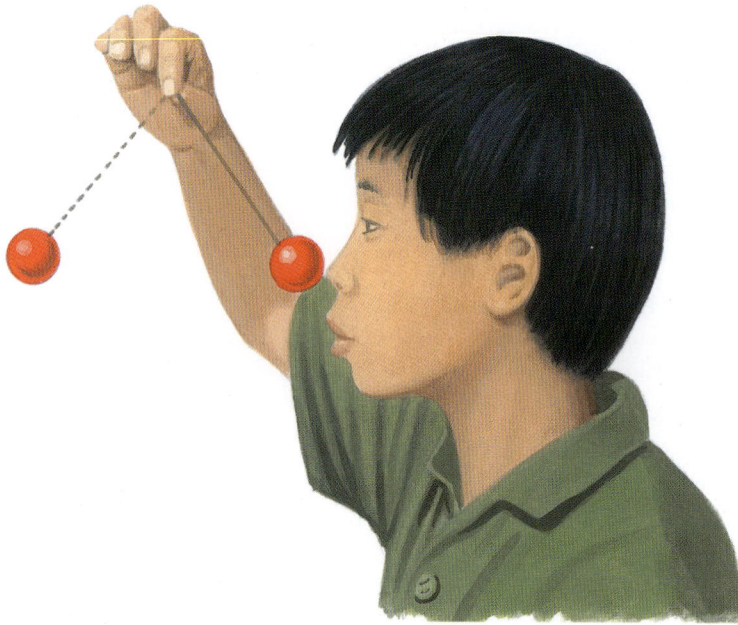

Figure 9.16 Using a pendulum.

15 What is the energy transfer diagram for the pendulum when it is
  a) released close to someone's nose,
  b) stops moving at the other end of its swing?

16 Why will the perpetual motion machine shown in the picture not run for ever?

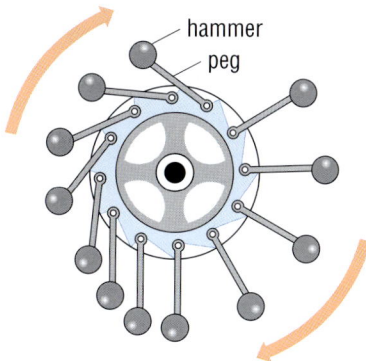

Figure 9.17 Perpetual motion machine.

This perpetual motion machine is called an overbalanced wheel. It is believed that someone in India thought of the idea for this machine about 1200 years ago. Since then many people in other parts of the world have worked on the idea but all have failed to make the machine work. The idea behind the machine is that the top of the wheel is given a push to the right. This makes the hammers on the right swing and fall. When they reach the bottom of the wheel they swing against a peg and push it to the left. If there is enough force in the push, the wheel should turn upwards on the left and this will cause more hammers on the right to fall. When models have been made, they work for a short while and then stop.

◆ SUMMARY ◆

- Energy transformations are energy changes (*see page 92*).
- When an energy transformation takes place, some energy is always lost as heat (thermal) energy (*see page 92*).
- Work and energy are measured in a unit called the joule (*see page 94*).
- Energy transfers can be shown by energy transfer diagrams which have three parts – energy input, energy converter and energy output (*see page 94*).
- Energy transfers can be shown by Sankey diagrams with arrows of different widths (*see page 95*).
- Seeds contain stored energy, which is used for germination and early growth of seedlings (*see page 95*).
- Some light energy falling on the leaves of a plant is converted into stored chemical energy inside the plant (*see page 96*).
- Chemical energy in food is released in respiration (*see page 97*).
- Electrical energy is easy to generate and transport (*see page 98*).
- A bicycle dynamo generates electricity (*see page 98*).
- Energy cannot be made or destroyed (*see page 100*).

End of chapter question

1 On your way home from school, make a list of examples of energy transformations. When you get home, draw an energy transformation diagram for each example.

10 Speeding up

Speed

Figure 10.1 How do you measure the runners' speeds?

A few moments after the finish of this race the times of the runners will be flashed up on a score board. If the race was over 100 metres, or any other distance, you may think that the time shows the speed that the runner ran that entire distance. However you would be wrong. The runners were stationary when the starter gun fired so they began to run fast or accelerate to get moving. They may have run steadily for most of the race then accelerated as much as they could for the final sprint to the finish. Speed is a steady rate of movement over a distance. To measure the speed of the runners they should have been running hard over the starting line and kept running steadily to the finish.

Speed records

Candidates trying to beat the land speed record must drive their car at full speed between two markers.

Figure 10.2 Breaking the land speed record in 1997.

Table 12.1 shows some land speed records from the end of the 20th century. You could check on the Internet to see if the 1997 record has been broken.

Table 10.1 Land speed records.

Date	Speed/km/h	Driver	Car
15/10/97	1227.99	Andy Green	Thrust SSC
4/10/83	1013.47	Richard Noble	Thrust 2
23/10/70	995.85	Gary Gabelick	The Blue Flame
15/11/65	960.96	Craig Breedlove	Spirit of America
7/11/65	922.48	Art Arfan	Green Monster

1 How long did Art Arfan's record stand?
2 How much faster than Spirit of America was The Blue Flame?
3 By how much did the land speed record rise between 1965 and 1997?

Ways of measuring speed

The speedometer

The speedometer in a car is connected by a cable to a shaft which turns the wheels. There is a wire in the cable which is connected to the shaft by gear wheels. When the shaft turns, the wire in the cable turns too. At the other end of the wire is a magnet. It spins round when the car wheels turn. The magnet is surrounded by a circular metal cup which is affected by the magnetic field generated by the spinning magnet. The cup is made to turn, the turning effect increasing as the speed of the spinning magnet (and the moving car) increases. The cup is connected to a spring and a pointer. The spring prevents the cup spinning but allows it to turn further as the car's speed increases. The pointer turns with the cup and moves across the scale of the speedometer dial.

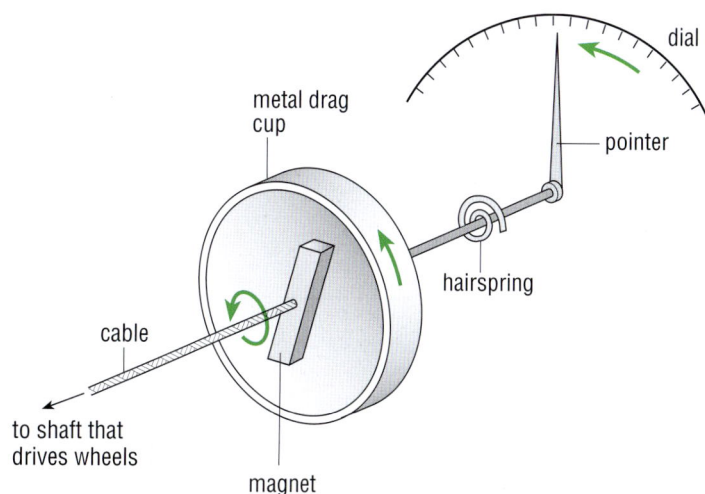

Figure 10.3 A speedometer.

The speed trap gun

The speed trap gun is a radar gun. When the gun is fired at an approaching vehicle a beam of radio waves travels to it through the air. This is reflected off the front of the vehicle and returns to a receiver on the gun. A computer in the gun compares the time difference between sending the beam and receiving it back from the vehicle and calculates the vehicle's speed.

The stop watch

For many years, the stop watch was used to measure speed. The watch was started as the speeding object passed the start line and was stopped when the object passed the finish line.

Light gates

In a light gate a beam of light shines onto a light-sensitive switch. The light gate used at the start of a speed test works in the following way. When the beam is broken by an object passing through it, the switch starts an electronic stopclock. The light gate used at the finish of the speed test causes the clock to be stopped when the beam is broken by the object passing through it.

4 Two people timed the speed of an object with a stop watch. They each got a slightly different result. How could this be?
5 Which is more reliable – using a manual stop watch or using light gates? Explain your answer.

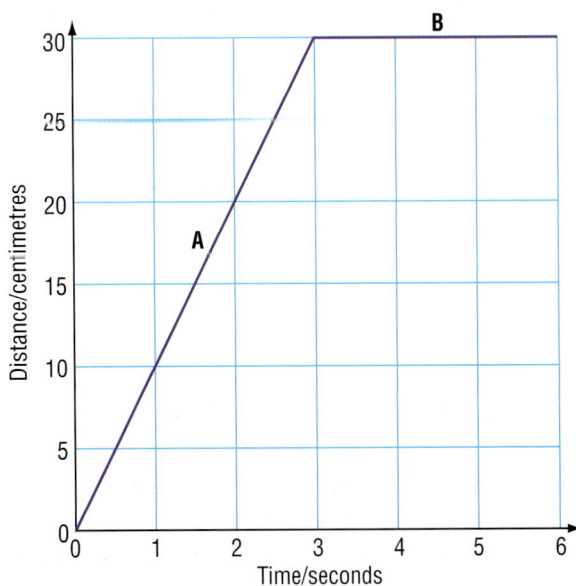

Figure 10.4 Distance/time graph.

Distance/time graph

The distance travelled by an object over a period of time can be plotted on a graph called a distance/time graph. The distance covered by the object is recorded on the Y axis and the time taken for the object to cover the distance is recorded on the X axis. When a distance/time graph is complete it can be used to study the speed of an object over different time periods of its journey.

Figure 10.4 shows the distance/time graph for an object which moved at a steady speed (line A) then stopped and remained stationary (line B). If the object had been travelling at a higher speed, line A would be steeper. If the object had been travelling at a lower speed, line A would be less steep.

The speed of an object, that is how far the object moved in a certain time, can be calculated from the distance/time graph. For example, the object in Figure 10.4 moved at 10 cm/s.

Velocity

When something moves it goes in a particular direction. An aeroplane, for example, may move at 900 km/h in a direction due north. The speed and the direction of movement when given together are called the velocity.

6 Figure 10.5 shows the distance/time graph for two trucks, A and B, on an expedition across the Mongolian desert.
  a) How far did truck A move in the first hour of its journey? What was its speed?
  b) How did the speed of truck A change in the second hour of its journey?
  c) Was truck B moving faster or slower than truck A in the first hour of its journey?
  d) What do you think might have happened to truck A in the third hour of the journey?

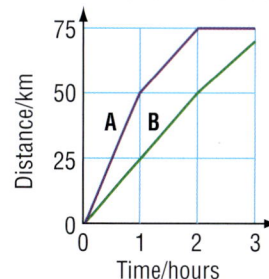

Figure 10.5

Acceleration

The acceleration of a moving object is a measure of how its velocity changes in a certain interval of time. The following equation shows how the steady acceleration of an object can be calculated:

$$\text{acceleration} = \frac{\text{change in velocity}}{\text{time}}$$

The SI unit for acceleration is m/s/s or m/s^2. This is pronounced metres per second per second or metres per second squared.

Friction

This contact force occurs between two objects when there is a push or a pull on one of the objects which could make it move over the surface of the other object. Friction acts to oppose that movement.

As the push or pull on the object increases, the force of friction between the surfaces of the objects also increases. This force matches the strength of the push or the pull up to a certain value and so, below this value, the object does not move. The friction which exists between the two objects when there is no movement is called static friction.

7 Imagine you are asked to push a heavy box across the floor. At first you need to push very hard but once the box has started to move you can push less strongly yet still keep it moving. Why is this?

8 When you take a step forwards you push backwards on the ground with your foot.

Figure 10.7

Make a sketch of Figure 10.7 and draw in an arrow to show the frictional force that stops your foot slipping.

9 Figure 10.8 shows a wheel turning.

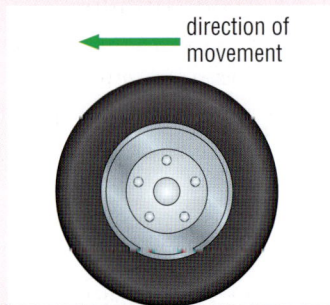

direction of movement

Figure 10.8

Make a sketch of the wheel. Draw and label

a) the force exerted by the wheel pushing backwards on the road, and

b) the force of friction preventing the wheel slipping.

If the strength of the push or pull on the object is increased beyond this value the object will start to slide. There is still a frictional force between the two surfaces, acting on each surface in the opposite direction to the direction of its movement. This frictional force is called sliding friction. The strength of this force is less than the maximum value of the static frictional force.

Figure 10.6 Friction between the log and the ground opposes the pulling force of the horse.

A closer look at friction

The surfaces of objects in contact are not completely smooth. Under a microscope it can be seen that they have tiny projections with hollows between them (Figure 10.9).

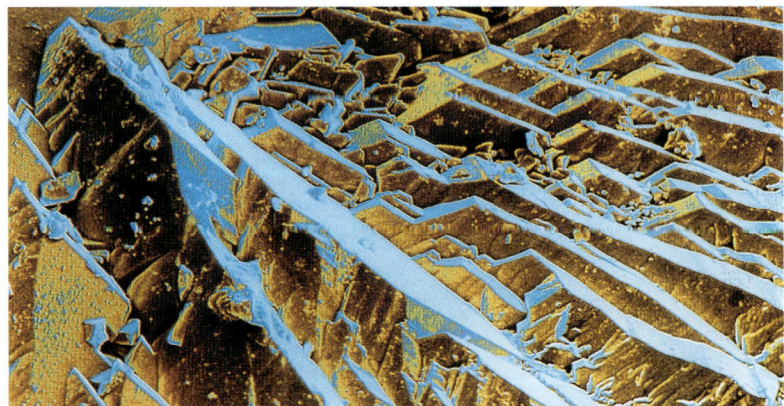

Figure 10.9 A metal surface that appears smooth to the naked eye has projections which can be seen when it is magnified 180 times.

Where the projections from the surface of one object meet the projections from the surface of the other, the materials in the projections stick. These connections between the surfaces produce the force of friction between the objects.

Reducing friction

If a liquid is placed between the two surfaces the projections are forced apart a little and the number of connections is reduced, which in turn reduces the force of friction. This can cause problems or it can be helpful. For example, water running between the surface of a tyre and the road reduces the friction between them and increases the chance of skidding. However, oil between the moving metal parts of an engine and the parts in the bearings reduces friction and also reduces wear on the metal parts.

10 Why does oiling the axles of a bicycle make the bicycle move more easily?

Increasing friction

The friction between two surfaces can be increased by pressing the surfaces together more strongly. This makes the projections press against each other more and increases the size and number of connections between the surfaces.

When brakes are applied on a bicycle or car the brake pads press against a moving part of the wheel and the force of friction increases. This opposes the rotation of the wheel and slows down the bicycle or car until it stops.

The tread on a car tyre is designed to move water out of the way as the tyre rolls over a wet road, reducing the risk of skidding. Racing cars have smooth tyres that are ideal for a dry track. If it rains they slide and skid all over the place and the tyres need to be changed.

Figure 10.10 Tyres with treads are designed so that water squirts out from between the treads.

Friction and road safety

When a driver in a moving car sees a hazard ahead the car travels a certain distance before the driver reacts and applies the brakes. The distance travelled by the car in this time is called the thinking distance. This is followed by the braking distance, which is the distance covered

11 a) What happens at the beginning of the time during which a car covers the thinking distance?

b) What happens at the end of the time during which the car covers the thinking distance?

12 What may affect the thinking time of the driver? How would the thinking distance of the car be affected? Explain your answer.

13 What, other than speed, may affect the braking distance of the car? Explain your answer.

14 A car is travelling along a road at 80 km/h when a tree falls across the road 54 metres away. What would probably happen and why?

by the car after the brakes are applied and before the car stops. Table 10.2 shows the thinking and braking distances that will bring a car with good brakes to a halt on a dry road.

Table 10.2

Speed	Thinking distance/m	Braking distance/m	Total stopping distance/m
48 km/h (30 mph)	9	14	23
80 km/h (50 mph)	15	38	53
112 km/h (70 mph)	21	75	96

For discussion

How safe is a) driving close to the car in front, b) driving fast on winding country roads with high hedges? Explain your answers to each part of the question.

'It's the driver that's dangerous not the car.'

Assess the usefulness of this slogan for a road safety campaign.

Other forces affecting speed

When objects move along a surface, friction occurs and opposes the motion. Two other forces that affect speed are air resistance and water resistance.

Air resistance

Air is a mixture of gases. When an object moves through the air it pushes the air out of the way and the air moves over the object's sides and pushes back on the object. This push on the object is called air resistance or drag.

The value of the air resistance depends on the size and shape of the object. Many cars are designed so that the air resistance is low when the car moves forwards. The car's body is designed like a wedge to cut its way through the air and the surfaces are curved to allow the air to flow over the sides with the

Figure 10.11 Testing a streamlined sports car in a wind tunnel.

minimum drag. Shapes that are designed to reduce air resistance are called streamlined shapes.

A dragster is a vehicle which accelerates very quickly. In a dragster race two vehicles accelerate along a straight track. At the end of the race the dragsters are slowed down by brakes and a parachute. The parachute offers a large surface area against which the air pushes. The high air resistance of the parachute slows down the dragster and helps it stop in a short distance.

Figure 10.12 These parachutes are used to slow down a dragster after a race.

15 How would the size of parachute required on a space probe to allow it to land safely differ on
a) a planet such as Venus which has a thick atmosphere, and
b) a planet such as Mars which has a thin atmosphere?
Explain your answers.

The air resistance produced by a parachute is also used to bring sky divers safely to the ground (see *Balanced forces*, page 112). The resistance of the gases in the atmospheres of other planets in the Solar System is used to slow down space probes so they can land safely and the devices on board are able to carry out their investigations.

Figure 10.13 A safe landing for the Mars Exploration Rover.

Water resistance

When an object moves through water it pushes the water out of the way and the water moves over the object's sides and pushes back on the object. This push on the object is called water resistance or drag. Objects that can move through the water quickly have a streamlined shape (see page 108). A fish such as a barracuda which moves quickly through the water has a much more streamlined shape than a slow-moving sunfish.

Barracuda.

Figure 10.14

A sunfish.

Water resistance affects the movement of ships and boats on the water surface. Boats designed for high speeds have a hull shaped to reduce water resistance as much as possible. Some boats are equipped with a device called a hydrofoil which reduces the area of contact between the boat and the water so that water resistance is kept to a minimum. The boat (itself called a hydrofoil) can then move quickly over the water surface.

A speed boat.

Figure 10.15

A hydrofoil.

16 Figure 10.17 shows a speed/time graph of a sky diver's jump.

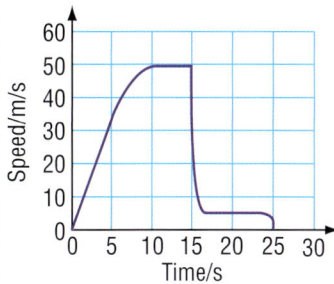

Figure 10.17

a) At what time is the acceleration greatest?

b) When did the sky diver begin to fall at the terminal velocity?

c) For how long did the sky diver fall at the terminal velocity?

d) When was the parachute opened?

e) At what speed did the sky diver hit the ground?

For discussion

During free fall, how does a sky diver alter the terminal velocity by altering his or her shape?

Why are tangled parachutes dangerous?

Sky diving: terminal velocity

When a sky diver leaps from an aeroplane the diver's weight pulls him or her down. The air resistance is small compared to the weight as he or she starts to fall and so the diver accelerates (Figure 10.16a). Eventually the force of the air resistance balances the force pulling the diver towards the ground and he or she falls steadily at what is called the terminal velocity (Figure 10.16b). When the sky diver opens the parachute the air resistance greatly increases (Figure 10.16c). This slows down the sky diver to a new, slower terminal velocity so he or she can make a safe landing.

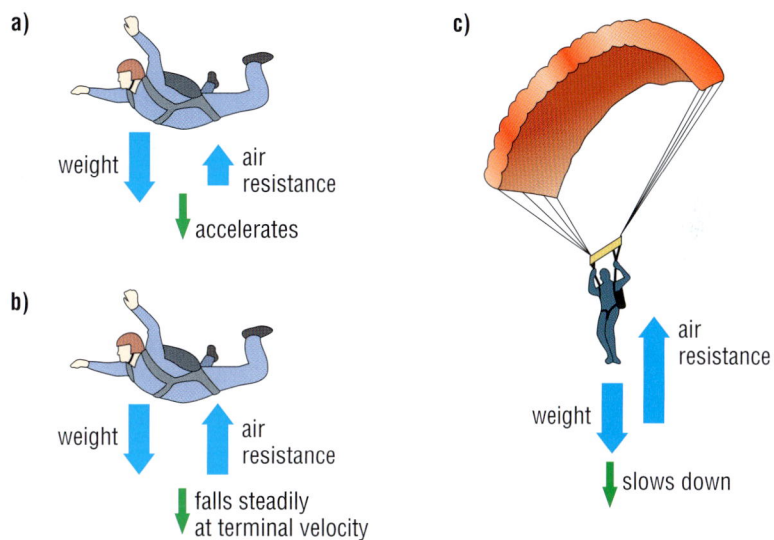

Figure 10.16 The motion of a sky diver.

Pairs of forces
Action and reaction

A force exerted by one object on another is always accompanied by a force equal to it acting in the opposite direction. For example, if you lean against a wall you exert a contact force that pushes on the wall and the wall exerts a contact force that pushes on you (Figure 10.18). The forces are equal. The force that you exert on the wall and the force the wall exerts on you are called an action–reaction pair. The action and reaction forces that form the pair act simultaneously: one does not cause the other.

Two model railway trucks each with a spring can be used to demonstrate an action–reaction pair. When the two trucks are pushed together the springs exert a force on each other due to the strain forces (see page 23)

111

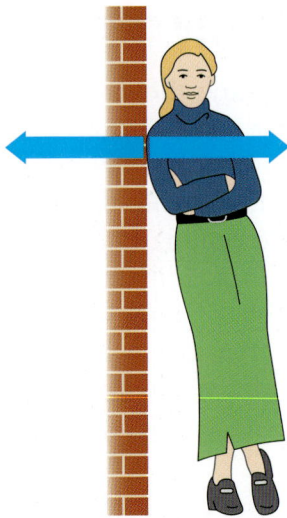

Figure 10.18 An action–reaction pair.

which develop in them. The force spring A exerts on spring B is always the same size but opposite in direction to the force exerted by spring B on spring A (Figure 10.19). The action force of A on B pushes B while the reaction force of B on A pushes A the other way. When the two trucks are suddenly released the action–reaction pair between the springs pushes the trucks in opposite directions.

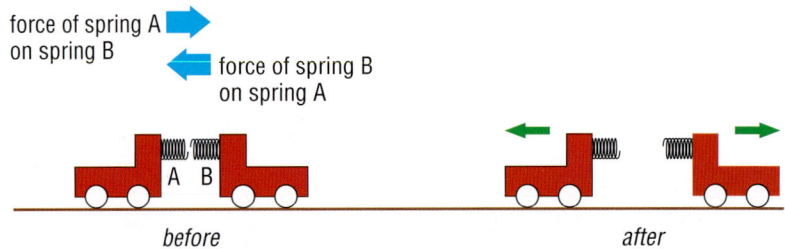

force of spring A on spring B

force of spring B on spring A

before   after

Figure 10.19 Action and reaction forces act in opposite directions.

Balanced forces

A pair of balanced forces is different from the action–reaction pair above, because balanced forces act on one object only. When you stand still you do not rise above the ground or sink into it, because of the two balanced forces acting on you. Your weight acts downwards into the ground because of gravity and the ground exerts a contact force upwards on the soles of your shoes. This contact force is equal to your weight.

A person sitting in a stationary go-kart does not move up or down because the weight of the person and the kart is balanced by the contact force of the ground on the tyres. When the kart is moving in a straight line at a constant speed (Figure 10.20) there is another pair of

17 Describe the forces between you and a chair when you sit on it.

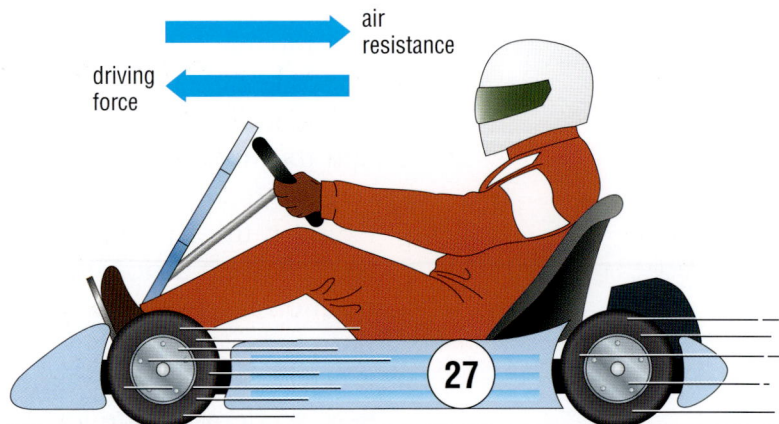

air resistance

driving force

Figure 10.20 Balanced forces act on the go-kart when it is moving at a steady speed.

balanced forces acting on the kart. These forces are the driving force pushing the kart forwards and air resistance pushing backwards on the kart.

Upthrust

When an object is placed in any liquid or gas it pushes some of the liquid or gas out of the way. The liquid or gas pushes back on the object with a force called the upthrust. This force is equal to the weight of the liquid or gas that has been pushed out of the way.

The buoy in Figure 10.21 is carrying a weather station. It floats on the sea surface due to the upthrust acting on it from the sea water.

18 What are the forces acting on a duck when it floats in water?

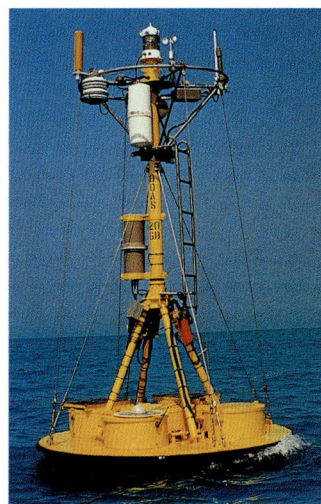

Figure 10.21 This floating buoy carries an automatic weather station. Its weight is balanced by the upthrust from the sea water.

Plimsoll lines

In the 19th century trade between countries increased and many kinds of goods were transported by sea. Some ships were loaded with so much cargo that they could barely float and when they encountered stormy conditions they sank. To reduce the number of shipping disasters and to save the lives of the sailors, the politician Samuel Plimsoll (1824–1898) brought in a law in 1876 requiring each ship to be marked with lines showing the maximum level to which the water should rise up its sides in dock. This prevented any ship from being overloaded and reduced the chance of it sinking in a storm. The lines became known as Plimsoll lines.

The symbol on the right in Figure A shows the loading limit in different situations. It shows how the level of the water will change as the ship moves into different kinds of water. These changes in level are due to the different densities of the different kinds of water, which leads to differences in upthrust. For example, the upthrust on a fully loaded ship in cold, dense sea water in the North Atlantic Ocean in winter is so strong that the ship only sinks to the level marked by WNA. If the fully loaded ship sailed into a warm, tropical, freshwater river it would sink to the level marked by TF.

1 What effect does the temperature of the water have on the upthrust of the water on a ship?

2 How does the presence of salt in water affect the upthrust of the water on a ship?

3 What might happen to a ship that was loaded to the line in cold sea water (W) and then sailed into warm fresh water? Explain your answer.

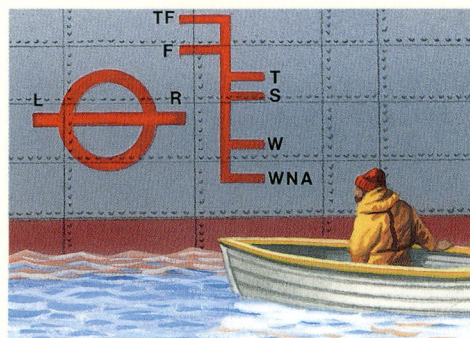

Figure A Plimsoll lines.

Unbalanced forces

When the forces on a stationary object are unbalanced the object starts to move. For example, when the driver of a go-kart presses the accelerator pedal on the stationary kart the wheels connected to the engine turn and the frictional force between the ground and the tyres pushes the kart forwards. As the kart moves forwards the air pushes on it (and the driver) with a force called drag or air resistance. To begin with this force is smaller than the frictional force and the kart continues to accelerate forwards (Figure 10.22).

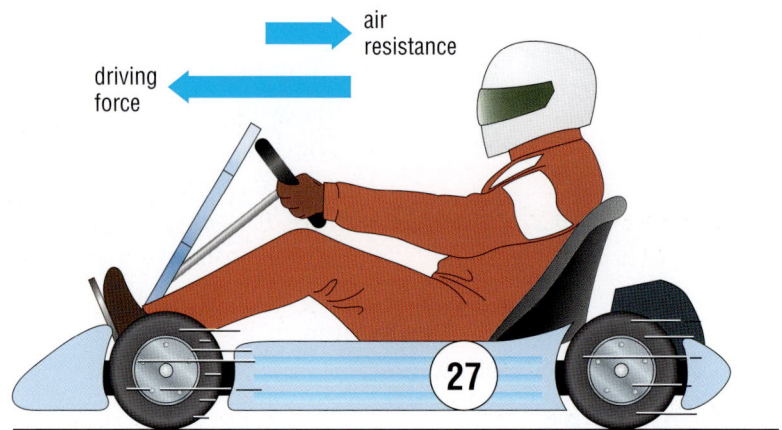

Figure 10.22   The forward force is greater than the backward force so the kart speeds up.

Note that the size of a force on a diagram is indicated by the size of the arrow. A large force is shown by a longer arrow than a small force.

As the kart increases speed the air resistance also increases. Eventually the kart moves at a constant speed in a straight line, as shown in Figure 10.20, with the two horizontal forces balanced.

When the driver takes his or her foot off the accelerator pedal the frictional driving force is reduced. The air resistance is now stronger than the driving force and the kart slows down.

The driver can slow down the kart faster by applying the brakes. The brake pads exert a frictional force on the wheels which makes it harder for the wheels to turn. This produces an additional resistance backwards which slows the kart down.

19 How does friction help you ride your bike?

20 Draw a submarine sinking in water. Draw and label the forces acting on it and indicate the strength of each one by the size of its arrow.

Mass and weight

21 What is the weight of the following masses on Earth:
a) 2 kg,
b) 3.5 kg,
c) 5.25 kg?

The mass of an object is a measure of the amount of matter in it. The weight of an object is the pull of the Earth's gravity on the object. For example, an object may have a mass of 1 kg. The pull of the Earth's gravity on 1 kg is a force of almost 10 newtons (actually 9.8 N but it is often rounded up to make the calculations easier). The weight of the 1 kg mass is therefore 10 N.

The region in which a force acts is called a field. There is a gravitational field around the Earth. The gravitational field strength is calculated by the equation:

$$\text{gravitational field strength} = \frac{\text{weight}}{\text{mass}}$$

At the Earth's surface we have seen that the pull on a mass of 1 kg is 10 N so the gravitational field strength is 10 N/kg.

The gravitational field strength on the surface of Mars is three times less than the gravitational field strength on the surface of the Earth. This means that a 1 kg object that is part of a space probe would have a weight of 10 N when it was on Earth but a weight of only 3.3 N on the surface of Mars.

The mass of an object remains the same wherever it goes in the Universe but its weight changes according to the gravitational force that is acting upon it.

22 What is the weight of a 6 kg object on the surface of Mars?
23 It is planned to bring samples of Mars rock back to the Earth. If 50 kg samples were collected by a space probe robot, what would be the weight of the rocks on
a) Mars,
b) Earth?
24 The Moon's gravitational field strength is one-sixth that of the Earth. What would be the weight of a 1 kg object on the Moon?
25 A sample of Moon rock weighed 30 N on the Moon.
a) What would be its weight when it was brought to the Earth?
b) What is the mass of the sample?

Weightlessness

The gravitational field strength around a planet, moon or star gets weaker and weaker as you move further away.

A space station in orbit above the Earth is still in the Earth's gravitational field. The force of gravity pulls on the space station but because the space station is moving with a velocity parallel to the surface of the planet, the force pulls the space station so that its path is curved. The force is just enough to keep the space station at exactly the same height. It does not move closer to Earth but 'falls' in a circular path around it. Inside the spacecraft every object that is not held down floats about. These objects, including the astronauts, are also 'falling' around the planet in the same way as the space station. The floating state is called apparent weightlessness because it feels like having no weight but the objects are, in fact, still being pulled by the Earth's gravity.

You may feel something similar to this weightlessness for a moment when you begin moving downwards in a lift or travel on a ride at a fair where you fall directly downwards. Both you and the ride are falling so you briefly feel lighter than usual. You may feel heavier just as the ride stops.

True weightlessness could only occur far out in deep space where there are no large objects with gravitational fields. This is beyond the distance travelled by any space exploration undertaken so far.

Figure 10.23 These people are enjoying a fairground ride where motion affects their apparent weight.

Gravity and space

We have all known about the force of gravity from our earliest days, as the force which makes things fall. It was introduced in the previous section as the gravitational force. In addition the relationship between mass and weight was examined and it was explained that the weight of an object is due to the pull of the Earth's gravitational force on that object.

Gravity and weight

26 The gravitational field strength on the Moon is one-sixth that of the Earth. If you lived on the Moon how would the weight of some everyday objects such as your watch, school bag or sports shoes change?

Although we may think of gravity as the force that pulls things down to the ground, it does more than that. It pulls them towards the centre of the Earth.

Objects released by an astronaut on the Moon fell to the Moon's surface. This indicates that the Moon also has a gravitational force. However, if an object is weighed on Earth and then weighed again on the Moon, its weight will be seen to have decreased. The decrease in weight is due to the less powerful gravitational force on the Moon. This is due to the mass of the Moon being much less than the mass of the Earth. The gravitational force of an object depends upon its mass.

Table 10.3 shows how the force of gravity exerted at the surfaces of different planets compare. The Earth's gravitational field strength is taken as 1 – the standard by which the others are compared.

Table 10.3 The planets' gravitational field strengths compared.

Planet	Gravitational field strength
Mercury	0.38
Venus	0.9
Earth	1
Mars	0.30
Jupiter	2.64
Saturn	0.925
Uranus	0.79
Neptune	1.12
Pluto	0.05

27 What would be the weights of the everyday things you considered in question 26 on
a) Mercury,
b) Jupiter,
c) Pluto?

Circular motion

A spinning object

At some time you have probably spun something around on a string. It may have been a key, a toy or a conker. The chances are you did not consider what was happening but you just enjoyed the spinning effect.

The object on the string was moving – so it possessed speed. However, at every moment its direction was changing. On page 105 velocity is described as the speed and direction of an object. This means that the velocity of a spinning object is always changing. Also, on page 105 acceleration is described as a measure of how an object's velocity changes in a certain time interval. This leads to the conclusion that an object moving in a circle is accelerating.

We usually think of acceleration as an increase in speed in a straight line. It occurs when a force pushes or pulls on an object to make it go faster. We have seen now that acceleration can also cause an object to move in a circle – but where is the force that causes this? For an object being spun on a string, the force is provided by the tension in the string. The force acts towards the centre of the circle, and is called a centripetal force.

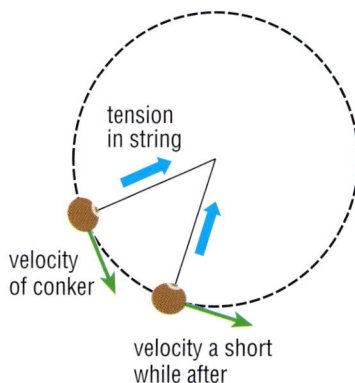

Figure 10.24 Circular motion.

28 a) If the string was cut while the object was spinning, in which direction would the object travel? Explain your answer.
b) How could you check your answer by experiment?

Moving the Earth

You are on an object which is moving through space on a curved path. Its velocity is constantly changing so an acceleration is taking place. The force producing the acceleration is a centripetal force produced by the gravitational force between the Sun and the Earth.

Gravitational forces exist between the Sun and the other planets, the asteroids and the comets, and cause them to move in their orbits in the Solar System.

How moons move

A moon is a large object which moves in an orbit around a planet. The force which causes this movement is the gravitational force between the planet and the moon.

Jupiter's gravity

It is believed that the gravitational force of Jupiter affects conditions inside two of its moons and causes warming. On Io this warmth is believed to have led to volcanic activity on the moon's surface. On Europa, an ice-covered moon, the warming is thought to have produced an ocean of water beneath the moon's surface. It is thought there could possibly be life there.

The force of Jupiter's gravity affects other objects in the Solar System. It is believed to have prevented the asteroids from joining together and forming a planet. It can also attract passing comets. In 1994 the comet Shoemaker-Levy 9 was trapped by Jupiter's gravity, broke up and fell into the planet's surface.

The strength of Jupiter's gravity even causes the Sun to wobble as it spins. This discovery of how a star can wobble in this way has led to the location of large planets around other stars.

> **For discussion**
> It has been said that Jupiter's presence has helped life to develop on Earth. Do you agree?

Figure 10.25 Jupiter as seen from Io, where its gravitational effect powers volcanoes.

◆ SUMMARY ◆

♦ Speed is a steady rate of movement over a distance (*see page 102*).

♦ Speed can be measured by using a speedometer, speed trap gun, stop watch or light gate (*see pages 103–104*).

♦ A distance–time graph shows the distance travelled by an object over a period of time (*see page 104*).

♦ The velocity of an object is its speed and its direction of movement (*see page 105*).

♦ The acceleration of an object is its change in velocity in a certain time interval (*see page 105*).

♦ Friction is a contact force which acts to oppose movement (*see page 105*).

♦ The push of the air on a moving object is called air resistance (*see page 108*).

♦ The push of water on an object moving through water is called water resistance (*see page 110*).

♦ When the forces on a falling object balance, it falls at its terminal velocity (*see page 111*).

♦ Forces act in pairs (*see page 111*).

♦ Forces on an object may be balanced (*see page 112*).

♦ Forces on an object may be unbalanced (*see page 114*).

♦ The mass of an object is a measure of the amount of matter in it (*see page 115*).

♦ The weight of an object is the pull of the Earth's gravity on the object (*see page 115*).

♦ A centripetal force makes one object move in a circular path around another (*see page 117*).

End of chapter questions

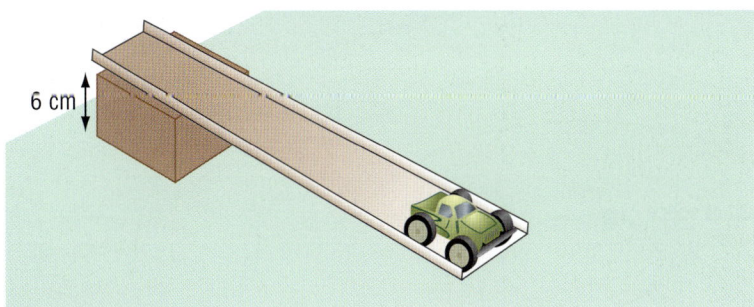

6 cm

Figure 10.26

A group of pupils investigated the movement of a model car. They set up a ramp at 6 cm height and let the car roll down it and across the floor (Figure 10.26). They measured the distance travelled by the car after it left the ramp and moved across the floor. The experiment was repeated three more times with the ramp set at 6 cm then the height was reset and more of the car's movements were recorded. Table 10.4 shows the results of the investigation.

Table 10.4

Height/cm	Distance/cm			
6	20	21	20	19
7	24	25	22	21
8	32	32	33	33
9	40	40	39.5	38
10	45	42	45	44
11	55	53	55	55
12	60	60	58	59
13	67	62	63	64

1 a) How many times was the height of the ramp changed?

  b) How is an average calculated?

  c) Calculate the average distance travelled for each height of the ramp.

  d) Plot a graph to show the relationship between the height of the ramp and the distance travelled from the ramp by the car.

  e) What conclusions can you draw from your analysis of the results of this investigation?

2 How do artificial satellites help people living on the Earth?

3 Some space probes have moved across the Solar System using the 'sling shot method' as shown in Figure 10.27.

  a) How do you think this method works?

  b) How do you think it saves on fuel?

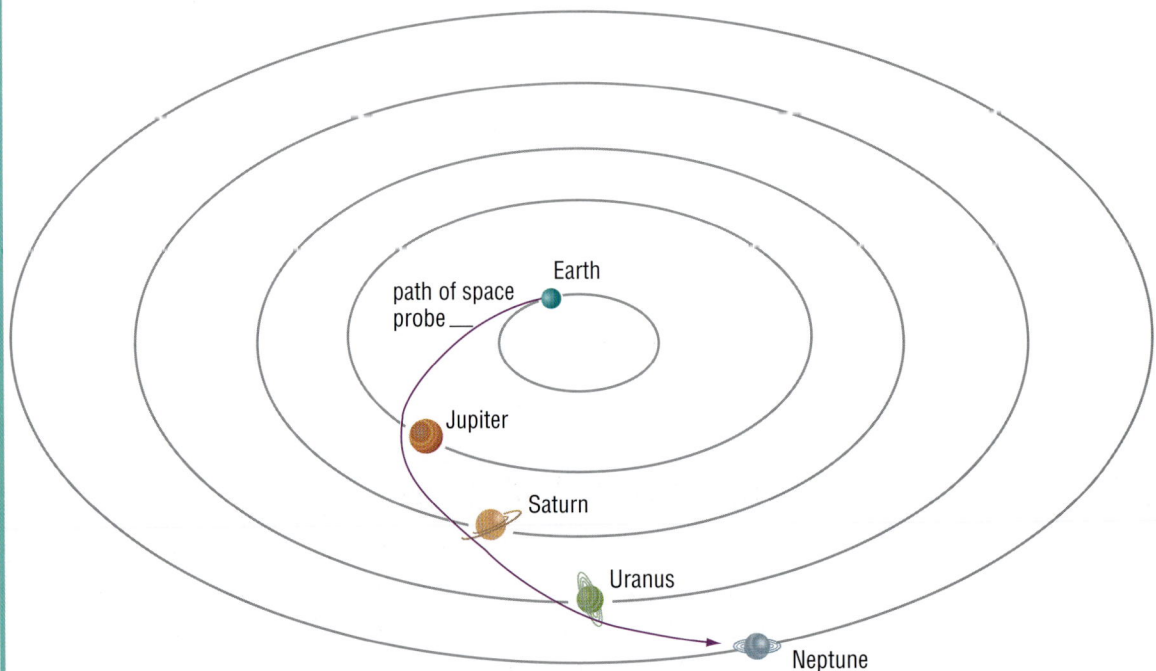

Figure 10.27   Sling shot method of sending a probe through the Solar System.

11

Electric charge

You may have seen a balloon stuck to the wall or ceiling, or received a slight shock when touching a car door, or heard a crackle when taking off your jumper.

To understand any of these you have to think about the structure of the atom and the electric charges on the particles in it.

The structure of the atom

An atom has a central nucleus surrounded by electrons. Each electron carries a negative electric charge (see also *Checkpoint Chemistry*, page 162). In the nucleus are particles called protons. Each proton carries a positive electric charge. Usually the number of positive charges carried by the protons is balanced by the number of negative charges carried by the electrons. For example, if an atom has six protons in its nucleus it has six electrons orbiting the nucleus. When the positive charges on the protons are balanced by the negative charges on the electrons in this way the atom is described as being neutral.

As you can see in Figure 11.2, there are also particles called neutrons in the nucleus. They have no electric charge.

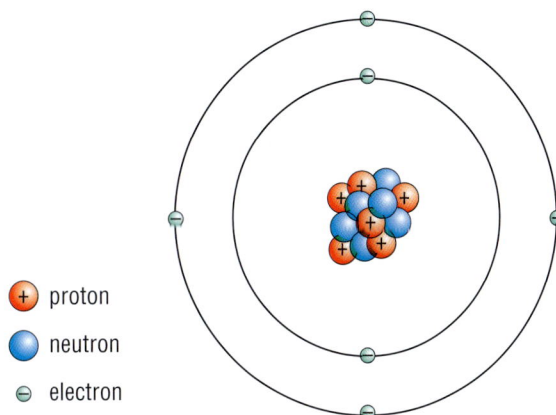

Figure 11.1 These balloons are held in place by electrostatic forces.

⊕ proton

🔵 neutron

⊖ electron

Figure 11.2 The structure of an atom.

Charging materials

In the party trick with the balloon, it must be rubbed on clothing, such as a woollen sleeve, before it will stick to the wall. When some dry materials are rubbed in this way they gain electrons from the atoms in the material they are being rubbed against. Other materials lose

1 When a piece of polythene is rubbed with a dry woollen cloth, electrons move from the cloth to the polythene. Which material becomes
   a) positively charged, and which
   b) negatively charged?

2 When a piece of perspex is rubbed with a dry woollen cloth, electrons move to the cloth. Which material becomes
   a) positively charged, and which
   b) negatively charged?

3 If a charged piece of polythene is set up as shown in Figure 11.4 and a charged piece of perspex is brought close to it, will the polythene swing towards the perspex or away from it? Explain your answer.

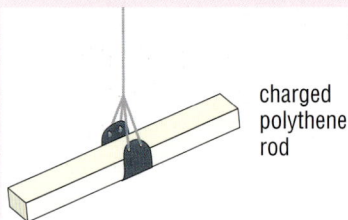

charged polythene rod

Figure 11.4

4 If a charged strip of polythene is set up as shown in Figure 11.4 and a charged polythene rod is brought close to it, will the polythene strip swing towards the polythene rod or away from it? Explain your answer.

electrons to the material they are being rubbed against. It depends upon the particular pair of materials involved. When a material that is an electrical insulator (see page 40) gains or loses electrons in this way, it is left with excess charge and the charge stays in place when the materials are separated. The material has been charged with static electricity.

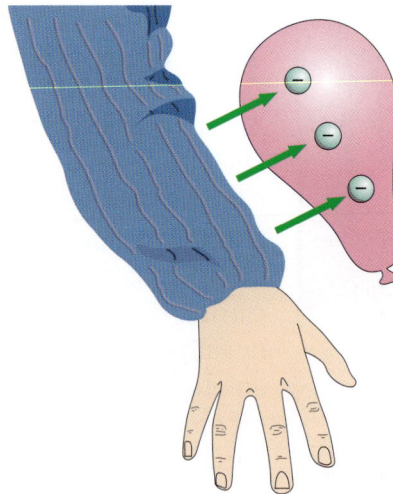

Figure 11.3 Electrons are transferred from the wool to the balloon and stay there.

A material that gains electrons when it is rubbed has more negative charges than positive charges and so is said to be negatively charged. A material that loses electrons when it is rubbed has fewer negative charges than positive charges and so is said to be positively charged. Protons are never transferred in this charging process since they are effectively locked in place in the nuclei of the atoms of the material.

When the balloon is rubbed on a sleeve it receives electrons from the material in the sleeve and its surface becomes negatively charged. If two balloons, suspended on nylon threads, are charged and placed close to each other they move apart. The negative charges on the balloons repel each other.

If a positively charged material is placed close to a negatively charged balloon hanging on a nylon thread the balloon moves towards the material because the different charges on the two materials attract each other (Figure 11.5).

5 a) When long dry hair is brushed the strands often move away from each other (Figure 11.6). Why do you think this happens?

b) The strands of hair also get attracted to the brush. Why do you think this happens?

Figure 11.6

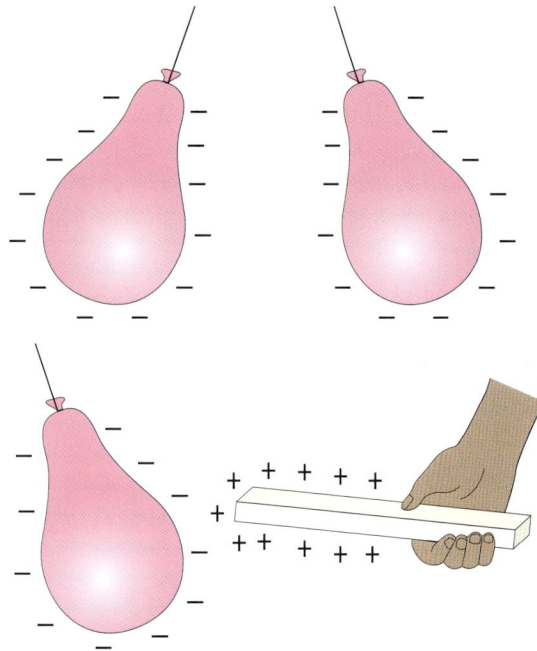

Figure 11.5   Similar charges repel (*top*) and different charges attract (*bottom*).

Early studies of electricity

For millions of years there have been certain kinds of trees which produce sap that turns to a clear yellow fossilised substance called amber. In Ancient Greece amber was used in items of jewellery. Thales (624–546BC), the earliest Greek 'scientific' philosopher, noticed that if amber was rubbed it developed the power to pick up small objects like dust, straw and feathers. The Greek word for amber is elektron so, much later, its attractive power became known as electricity.

William Gilbert (1544–1603), an English scientist, discovered that a few other materials, such as certain gemstones and rock crystal, could also attract small objects when they were rubbed. He called these materials 'electrics'.

1 Where did the word 'electricity' come from?

2 How did Gilbert extend the observations of Thales?

3 Why do you think electric machines amused people?

Figure A   Using an electric machine to ignite a sample of wine!

(continued)

Otto von Guericke (1602–1686) used an 'electric' called sulphur to make a machine that could generate sparks. He made the sulphur into a ball and attached it to an axle which could be turned quickly by a crank handle. As the ball spun it was rubbed and built up a charge of static electricity which produced sparks. Electric machines became popular as a form of amusement and entertainment. Some people made their living by travelling through European countries, demonstrating their machines.

Stephen Gray (1696–1736), another English scientist, investigated electricity by rubbing a glass tube that had corks in the ends, and discovered that the corks became charged with static electricity even though they had not been touched. He had discovered that electricity behaved as if it could flow like a liquid.

Charles Du Fay (1698–1739), a French scientist, repeated Gray's experiments and extended them by comparing the way in which the objects were charged with electricity. He discovered that if he charged a cork ball using a glass rod that had been rubbed, it was attracted to a cork ball that had been charged using sealing wax that had been rubbed. He also discovered that two cork balls charged by either the rubbed glass or sealing wax repelled each other. From his investigations he believed that electricity was made from two different liquids. They were called 'vitreous electricity' (from rubbing glass) and 'resinous electricity' (from rubbing sealing wax).

4 How was Gray's idea developed by Du Fay?
5 How was Du Fay's work developed by Franklin?
6 How is Franklin's idea of electric charge
   a) similar to, and
   b) different from
   the ideas that we use today?

Figure B   An electric machine built in about 1762.

Benjamin Franklin (1706–1790), an American statesman and scientist, refined Du Fay's idea of two electrical fluids by suggesting that when substances were charged they received either too much fluid and became positively charged or they had some fluid taken away and became negatively charged.

Insulators and conductors

A material that can become charged with static electricity is called an insulator. If electrons are added to the material they stay in place and the insulator is negatively charged. If electrons are removed from the material more electrons do not flow into the material and it remains positively charged.

A metal cannot be charged with electricity by rubbing in the way an insulating material can because electrons flow easily through metals. A material through which electrons can flow is called a conductor. The human body is a very good conductor of electricity.

Induced charges

If a material has an electric charge it can make or 'induce' an electric charge on the surface of a material close by without touching the material. For example, if a piece of plastic, such as a pen, is rubbed and held above a tiny piece of paper, the positive charge on the plastic draws electrons to the surface of the paper nearest the plastic. This makes the uppermost surface of the paper negatively charged. When the pen is brought very close to the paper the force of attraction between the two surfaces is strong enough to overcome the weight of the paper and the paper springs up to the surface of the pen.

The underside of the paper is left with a positive charge but since this is further away from the pen the force of repulsion it experiences is weaker than the attractive force and the paper is held.

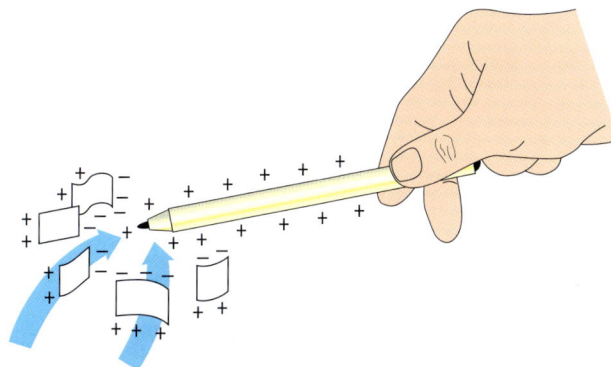

Figure 11.7 *The charged pen induces charges on the surfaces of the paper.*

In a similar way a charged balloon induces an opposite charge on the surface of a wall it is brought close to.

6 When a negative charge is
induced on one surface of a
piece of paper what is induced
on the other surface of the
paper? Why does this happen?
7 Why does a rubbed balloon stick
to the wall?

When the balloon touches the wall, the force of attraction between the two surfaces is greater than the weight of the balloon and the air inside it so the balloon sticks to the wall (see Figure 11.1).

This way of charging a material without touching it is called charging by induction.

Sparks and flashes

Air is a poor conductor of electricity but if the size of the charge on two oppositely charged surfaces is very large the air between them may conduct electricity as a spark or a flash, like a flash of lightning. This happens when the molecules in the air are split. They form negatively charged electrons and positively charged ions. The electrons move towards the positively charged surface and ions move towards the negative surface (Figure 11.8). As the electrons move they collide with other molecules in the air and split them. The ions and electrons from these molecules also move towards the charged surfaces and split more molecules as they go.

This process occurs very quickly and produces a spark. When the charged particles in the air meet the charged surfaces the positive and negative charges cancel out each other and the surfaces lose their charge. They are said to have been discharged.

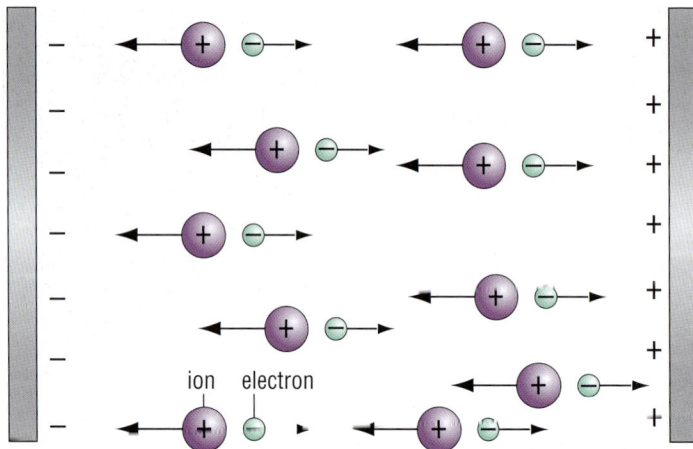

ion   electron

Figure 11.8 The strong electric field between the charged plates ionises the air between them.

Preventing an explosion

When an aircraft flies through the air its surfaces are rubbed by air particles and become charged with static electricity. If the aircraft was equipped with non-conducting tyres, such as those used on most vehicles, the charge would remain on the aircraft when it landed. This

Figure 11.9 Sparks must be prevented when an aircraft is being refuelled.

charge could cause a spark during refuelling. The heat from the spark would be sufficient to cause the fuel vapour to combust which would result in a devastating explosion. This danger is prevented by equipping the aircraft with tyres that conduct electricity. When the aircraft lands the charge it possesses passes to the ground.

Lightning

When a storm cloud develops, strong winds move upwards through the cloud and rub against large raindrops and hail stones. This rubbing causes the development of charged particles in the cloud. Positively charged particles collect at the top of the storm cloud and negatively charged particles collect at the base. The size of the different charges in each part of the cloud may become so large that lightning, called sheet lightning, is produced between them.

The negative charge at the base of the storm cloud induces a positive charge on the ground below. If the charges become large enough a flash of lightning, called forked lightning, occurs between them.

Figure 11.10   Forked lightning.

The van de Graaff generator

In 1931 Robert van de Graaff invented a machine which produced a huge charge of static electricity. The machine is called the van de Graaff generator.

In the generator is a rubber belt which runs over two rollers. One roller is made of perspex and the other is made of polythene. The belt is driven by an electric motor. When the perspex roller is placed at the base of the generator, the belt running over it becomes negatively charged (Figure 11.11). The charged part of the belt rises to the polythene roller at the top of the generator where there is a device that transfers the negative charge to the hollow metal dome. The belt moves over the roller in the dome and back to the roller at the base, where it becomes negatively charged again.

When the rollers are reversed and the polythene roller is placed at the base of the generator, the belt becomes positively charged and the positive charge is transferred to the dome.

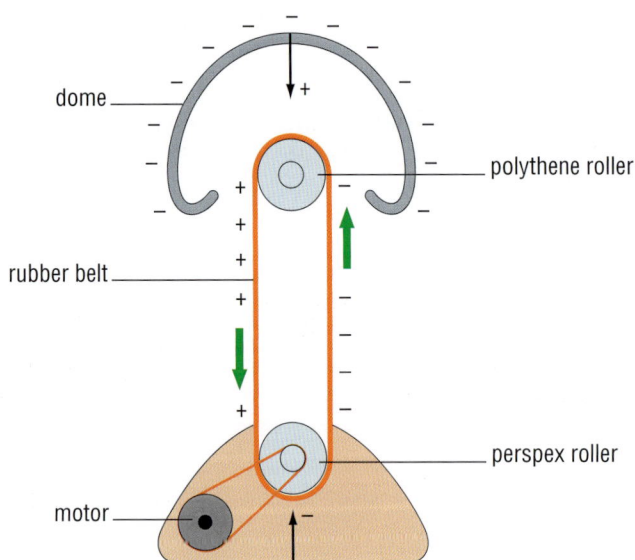

Figure 11.11 Inside a van de Graaff generator.

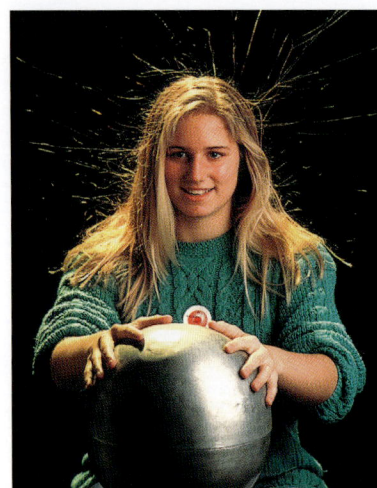

Figure 11.12 Demonstrating a large electrostatic charge with a van de Graaff generator.

Very high charges can be stored in the dome and released during investigations. In the past huge van de Graaff generators were used as particle accelerators to investigate atomic structure but they were replaced by other devices. Today they are used in schools and colleges to generate quite large electrostatic charges for demonstration lessons.

8 a) Explain what you see in Figure 11.12.
b) Why must the pupil be standing on a sheet of insulating material?

Measuring current

The rate at which electrons flow through a wire is measured in units called amperes. This word is usually shortened to amps and the symbol for it is A. One amp is equal to the flow of 6 million, million, million electrons passing any given point in the wire in a second!

The current flow in a circuit is measured using an instrument called an ammeter. This is a device which has a coil of wire set between the north and south poles of a magnet. The coil has a pointer attached to it and it turns when a current passes through it. The amount by which the coil turns depends on the size of the current and is shown by the movement of the pointer across the scale.

When an ammeter is used it is connected into a circuit with its positive or red terminal connected to a wire that leads towards the positive terminal of the cell, battery or power pack. It is always connected in series with the component through which the current flow is to be measured (Figure 11.13). Ammeters usually have a very low resistance so that the current passes through them without affecting the rest of the circuit.

9 How many electrons are flowing per second past a point in a circuit in which there is a current of
   a) 0.5 amps,
   b) 5 amps,
   c) 30 amps?
10 Towards which terminal of the power supply should the negative or black terminal of an ammeter be connected?

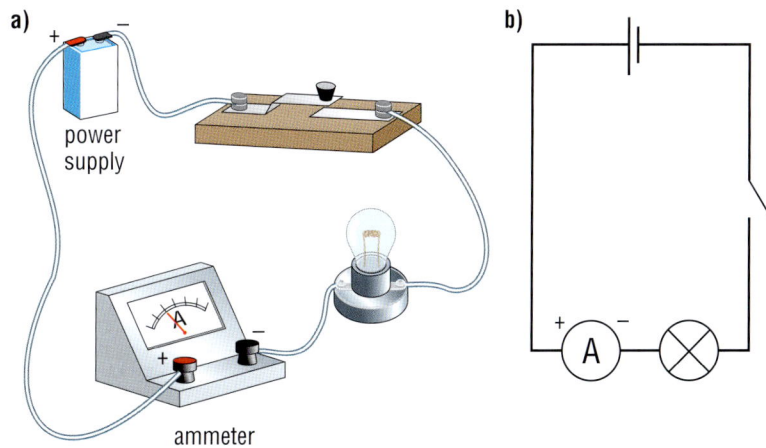

Figure 11.13 **a)** An ammeter connected in a circuit and **b)** the circuit diagram showing its symbol.

When an ammeter is to be used to measure the current flowing through a series circuit such as that shown in Figure 11.14a, the ammeter is placed at a position such as A or B.

When an ammeter is to be used to measure the current flowing through a parallel circuit such as that shown in Figure 11.14b, the ammeter should be placed at A, B and C in turn.

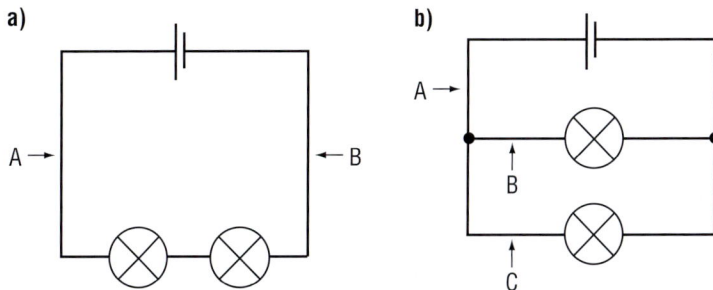

Figure 11.14 Measuring current in **a)** series and **b)** parallel circuits.

The current in a series circuit

In a series circuit as shown in Figure 11.15 the ammeter will record the same amount of current in each position A, B, C and D.

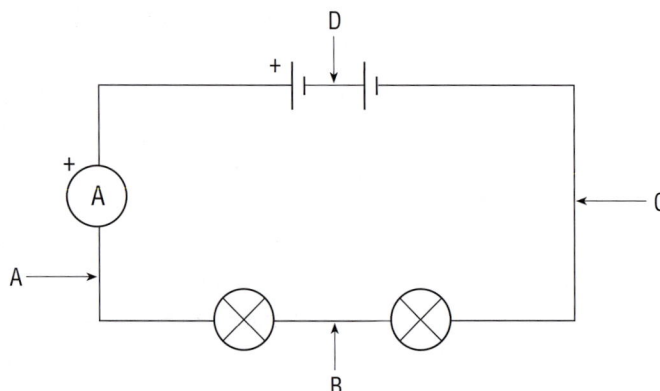

Figure 11.15 Measuring current in a series circuit using an ammeter.

◆ SUMMARY ◆

- ◆ An atom has two kinds of electrically charged particles. They are protons and electrons (*see page 121*).
- ◆ Static electricity may be generated by rubbing insulators (*see page 121*).
- ◆ Charged materials can have either a positive or a negative charge (*see page 122*).
- ◆ Materials that hold charges of static electricity are called insulators; materials that allow electricity to pass through them are called conductors (*see page 125*).
- ◆ A material may become charged by induction (*see page 125*).
- ◆ Static electricity can be discharged through the air as a spark or a flash (*see page 126*).
- ◆ The van de Graaff generator can produce large charges and may be used to study the effects created by static electricity (*see page 127*).
- ◆ An ammeter measures the rate of flow of electrons (current) (*see page 129*).

End of chapter question

1 How could you use your knowledge of the structure of the atom to explain how a plastic pen that has been rubbed can pick up tiny pieces of paper?

12 *Heat energy transfer*

Heat and internal energy

The 'heat' in a substance is really a measure of the total kinetic energy of the atoms and molecules of a substance, due to its internal energy. The total amount of heat in a substance is related to its mass. A large mass of a substance holds a larger amount of heat – it has more internal energy – than a smaller mass. For example, if $100\,cm^3$ of water is heated in a beaker with a Bunsen burner on a roaring flame it will take less time to reach $100\,°C$ than $200\,cm^3$ of water would because it has a smaller mass.

Figure 12.1   When heating two masses of water, more heat energy needs to be supplied to the larger mass to reach the same temperature.

When a substance is heated the (thermal) energy supplied increases the internal kinetic energy which means the atoms and molecules in the substance move faster and further. If the temperature of the substance is taken with a thermometer, kinetic energy from the substance passes to the atoms or molecules from which the thermometer liquid is made and causes them to move faster too. This leads to an expansion of the liquid in the thermometer tube. The thermometer measures the (average) kinetic energy of the particles hitting the bulb and not the total kinetic energy of all the particles in the substance.

1 Why does it take a full kettle longer to boil than a half-full kettle?

2 Which do you think contains more internal energy, a teaspoon of boiling water or a pan full of water at $50\,°C$?

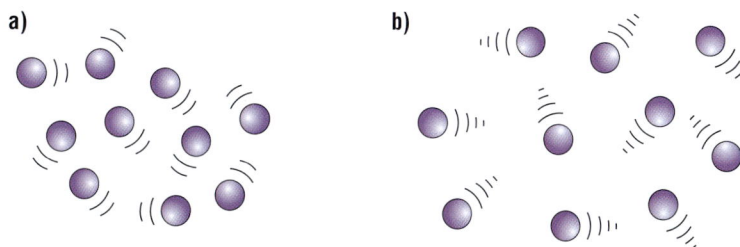

Figure 12.2   Particles in motion in **a)** a cool substance and **b)** a hot substance.

Measuring the amount of heat energy

The amount of heat (thermal) energy given to a substance can be measured by heating the substance with an electric heater. The quantity of electrical energy used can be measured by a joulemeter and this equals the amount of heat (thermal) energy supplied.

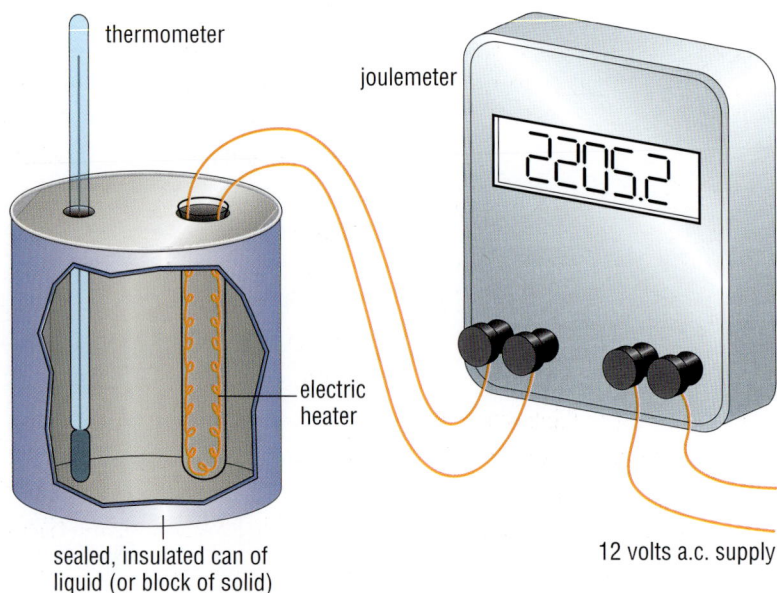

thermometer

joulemeter

2205.2

electric heater

sealed, insulated can of liquid (or block of solid)

12 volts a.c. supply

Figure 12.3 Measuring heat energy supplied with a joulemeter.

The equipment in Figure 12.3 can be used to compare how the heat supplied to a liquid or solid affects its temperature. It is found that some substances, such as water, take up large amounts of heat energy but their temperature only rises a few degrees, while the same mass of other substances needs only a small amount of heat energy to raise their temperature by the same amount.

How heat energy travels

There are three ways in which heat energy can travel. They are conduction, convection and radiation. Together they are known as thermal energy transfer.

Conduction

The heat energy is passed from one particle of a material to the next particle. For example, when a metal pan of water is put on a hot plate of a cooker the atoms in the metal close to the hot plate receive heat energy

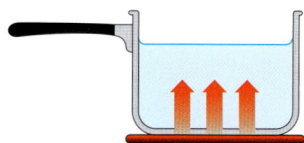

Figure 12.4 The conduction of heat through the bottom of a pan.

3 A metal rod had drawing pins stuck to it with wax and was heated at one end as shown in Figure 12.5.

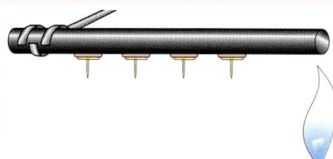

Figure 12.5

a) What do you think will happen in this experiment? Explain your answer.

b) How could this experiment be adapted to compare the conducting properties of different materials?

4 Imagine that a football represents heat energy and football players represent particles in a material. Which of the following events is like

a) conduction, and

b) convection?

Explain your answers.

i) The players pass the ball to each other to move it up the field.

ii) A defender receives the ball and runs up field with it into an attacking position.

5 When coal burns, particles of soot rise up above the fire and make smoke. Why doesn't the smoke move along the ground?

and vibrate more vigorously. They knock against the atoms a little further into the bottom of the pan and make them vibrate more strongly too. These atoms knock against other atoms a little further up and the kinetic energy is passed on. Eventually the inner surface of the pan, which is next to the water, becomes hot too.

Conduction can occur easily in solids, less easily in liquids but hardly at all in gases because the gas atoms are too far apart to affect each other. It cannot occur in a vacuum, such as outer space, where there are no particles to pass on the heat energy. Conduction is fastest in metals because they have electrons that are free to move. When a metal is heated the electrons in that part move about faster and pass on heat energy to nearby electrons and atoms, so that the heat energy spreads quickly through to cooler parts of the metal.

Materials that allow heat to pass through them easily are called conductors of heat. Materials that do not allow heat to pass through them easily are called insulators. Insulators are useful in reducing the loss of heat energy. For example, the fabric of a thick woollen pullover is a good insulator. It keeps in your body heat in cold conditions. Insulation materials are also used on the floor of lofts to reduce the escape of heat through the roofs of houses.

Convection

The heat energy is carried away by the particles of the material changing position. For example, the water next to the hot surface at the bottom of the pan receives heat from the metal. The molecules of water next to the metal move faster and further apart as their kinetic energy increases. This makes the water next to the pan bottom less dense than the water above it and the warm water rises. Cooler water from above moves in to take the place of the rising warmer water. The cool water is also warmed and rises. It is replaced by yet more cool water and convection currents are set up as shown in Figure 12.6.

Figure 12.6 The convection currents in a pan of water heated from below.

6 a) The temperature of the land
surface is higher than the
temperature of the sea
surface during the day. Use
the ideas of convection
currents to suggest what
happens to air above the
land and above the sea.
Which way do you think the
wind will blow across the
promenade in Figure 12.7?
Explain your answer.

Figure 12.7

b) At night the land surface is
cooler than the sea surface.
Does this affect the wind
direction? Explain your
answer.

Convection can only occur in liquids and gases. It
cannot occur in solids where the particles are not free to
move about, nor in a vacuum such as outer space.

Radiation

Energy can travel through air or through a vacuum as
electromagnetic waves (see page 30). For example, as
the pan of water gets hotter you can put your hand near
its side and feel the heat on your skin even though you
are not touching the metal. The sides of the pan are
radiating infrared waves. These carry the heat energy
from the surface of the pan to the surface of your skin,
which is warmed by them.

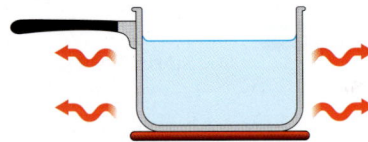

Figure 12.8 Heat radiation from a hot pan.

All objects radiate infrared, but the hotter the object the
more infrared energy it radiates and the shorter the
wavelength of the waves.

Some infrared radiation can pass through certain
solids such as glass. For example, the infrared radiation
from the Sun can pass through glass in a greenhouse but
the (longer wavelength) infrared radiation from the
ground and the plants inside the greenhouse cannot
pass back out through the glass. This infrared radiation
is trapped and warms the contents of the greenhouse.

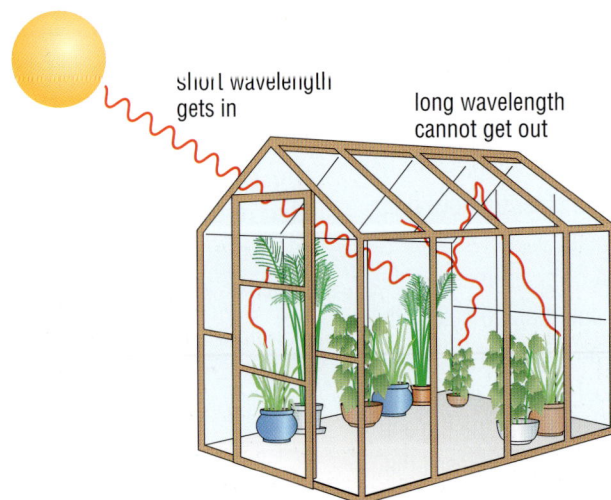

short wavelength
gets in

long wavelength
cannot get out

Figure 12.9 A greenhouse traps infrared energy.

7 In question 4 the energy was transferred by particles (players). How is the transfer of heat energy by radiation different?

8 How does the type of surface of an object affect the way it radiates and absorbs heat energy?

The type of surface affects the amount of heat energy radiated from an object in a given time. Darker colours radiate energy more rapidly than lighter colours, and black surfaces radiate the most rapidly. The surfaces which radiate the energy least rapidly are light shiny surfaces, like the surface of polished metal.

The type of surface also affects the amount of radiated energy absorbed by the surface in a given time. For example, a light shiny surface absorbs energy the least rapidly, while a dull black surface absorbs energy the most rapidly.

The Thermos flask

Sir James Dewar (1842–1923) studied how gases could be turned into liquids. Liquid oxygen has a boiling point of −182.9 °C. When the liquid oxygen was made it needed to be stored in a container which would prevent heat from the surroundings entering and causing the liquid to boil. In 1892 Dewar invented a flask called a Dewar flask which allowed him to keep the liquid oxygen cool for his experiments. The Dewar flask is now more widely known as the Thermos flask (Figure A) and is used mainly to keep drinks hot.

1 Which forms of energy transfer does the vacuum prevent?

2 Why are the glass walls shiny? How would the efficiency of the flask be affected if the walls were painted black? Explain your answer.

3 How could a warm liquid lose heat if the stopper was removed? Explain your answer.

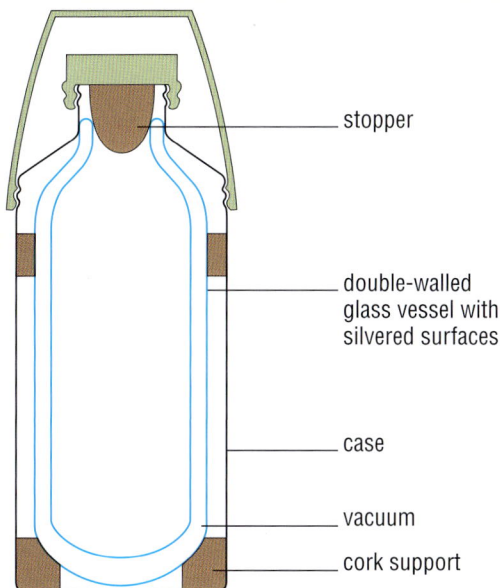

Figure A Structure of a Thermos flask.

stopper

double-walled glass vessel with silvered surfaces

case

vacuum

cork support

The walls of the flask are made of glass, which is a poor conductor of heat, and are separated by a vacuum. The glass walls themselves have shiny surfaces. The surface of the inner wall radiates very little heat and the surface of the outer wall absorbs very little of the heat that is radiated from the inner wall. The cork supports are poor conductors of heat and the stopper prevents heat being lost by convection and evaporation in the air above the surface of the liquid.

Evaporation

The particles in a liquid have different amounts of energy. The particles with the most energy move the fastest. High energy liquid particles near the surface move so fast that they can break through the surface and escape into the air and form a gas.

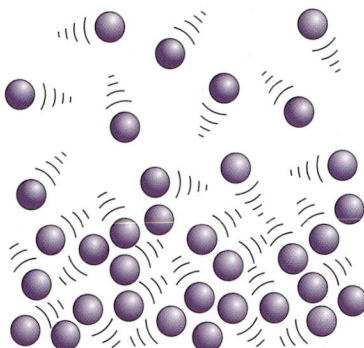

Figure 12.10 Evaporation.

◆ SUMMARY ◆

- ◆ When a substance is heated its internal energy increases and its particles move faster (*see page 131*).
- ◆ A thermometer measures the average kinetic energy of the particles hitting the thermometer bulb (*see page 131*).
- ◆ Heat energy supplied can be measured with a joulemeter (*see page 132*).
- ◆ Heat energy is passed from particle to particle by conduction (*see page 132*).
- ◆ Materials can be divided into conductors and insulators according to how easily heat passes through them (*see page 133*).
- ◆ Heat energy is carried by moving particles in a convection current (*see page 133*).
- ◆ Heat energy is carried by electromagnetic waves in radiation (*see page 134*).
- ◆ Evaporation can be explained by considering the amount of energy in the particles in a liquid (*see this page*).

End of chapter questions

An investigation was carried out to see if useful amounts of heat energy from the Sun could be trapped in trays of water.

   Three metal trays were used. Each one was 25 cm long, 20 cm wide and 5 cm deep and was filled with 1500 cm^3 of water.

Tray 1 had a glass plate cover and the water in it was untreated.
Tray 2 had some black ink added to the water before the glass plate cover was put over it.
Tray 3 had some black ink added to the water before the glass plate cover was put over it, then the sides and base were packed with vermiculite – a spongelike, rocky material.

The trays were exposed to sunlight during the day for 7 hours and the air temperature and the temperature of the water in each tray were taken every hour. Table 12.1 shows the data that were collected.

Table 12.1

Time	Air temperature/°C	Water temperatures/°C		
		colourless (without insulation)	black (without insulation)	black (with insulation)
11.15 am	18.7	17.5	17.5	17.5
12.15 pm	19.0	18.9	20.0	20.4
1.15 pm	20.5	22.4	25.1	26.0
2.15 pm	18.3	21.0	23.0	24.9
3.15 pm	18.9	21.5	23.4	25.6
4.15 pm	20.6	24.5	28.3	29.1
5.15 pm	20.2	28.5	31.8	33.2
6.15 pm	19.2	26.0	28.1	31.5

1 Did the water fill the trays to the top? Explain your answer.
2 Plot lines of the data for the air temperature and the temperature of each tray, on the same graph.
3 Compare the graphs you have drawn.
4 What was the purpose of the ink and the vermiculite? Explain your answers.
5 What is the maximum temperature rise? When was it achieved and in which tray?
6 What can you conclude from this investigation?

13 Pressure and density

Which is heavier – the wood in the trunk of a tree or the metal in a coin? Your first answer might be to say the trunk of the tree, but it can float on water while a coin would quickly sink to the bottom. To be a fair comparison we need to find the mass of equal volumes. If we found the mass of a piece of wood the size of a coin we would see it was lighter.

Figure 13.1 Transporting timber by water.

Density

The density of a substance is a measure of the amount of matter that is present in a certain volume of it. The following equation shows how the density of a substance can be calculated:

$$\text{density} = \frac{\text{mass}}{\text{volume}}$$

The basic SI unit of density is found by dividing the unit of mass by the unit of volume, so it is kg/m^3. This is pronounced kilograms per metre cubed.

Table 13.1 shows the densities of some common solid materials.

Table 13.1 The density of some common solid materials.

Material	Density/kg/m^3
ice	920
cork	250
wood	650
steel	7 900
aluminium	2 700
copper	8 040
lead	11 350
gold	19 320
polythene	920
perspex	1 200
expanded polystyrene	15

1 Arrange the materials in Table 13.1 in order of density, starting with the least dense material.
2 Which is heavier, a cubic metre of steel or a cubic metre of aluminium?
3 Which is heavier, a kilogram of steel or a kilogram of cork?

In the school laboratory when small amounts of materials are used the density of a substance is often calculated using masses measured in grams and volumes in cubic centimetres, giving a density value in g/cm^3. The density value in units of g/cm^3 can be converted to a value in kg/m^3 by multiplying it by 1000.

4 A block of material is 8 cm long, 2 cm wide and 3 cm high, and has a mass of 46 g. What is its density?

5 a) Convert the density value you found in question 4 to kg/m^3.

b) Compare the density of the material in the block with those in Table 13.1. Which materials in the table have densities closest to that of the block?

c) How could you convert the value of a density given in kg/m^3 to g/cm^3?

For example, ice was found to have a density of 0.920 g/cm^3. This can also be expressed as 0.920 × 1000 = 920 kg/m^3.

Measuring the density of a rectangular solid block

The mass of the block is found by placing the block on a balance (check the balance reads zero first) and reading the scale. The mass in grams is recorded. The volume is found by multiplying the length, width and height of the block together and recording the value in cubic centimetres. The density of the material in the block is found by dividing the mass by the volume and expressing the quantity in the unit g/cm^3.

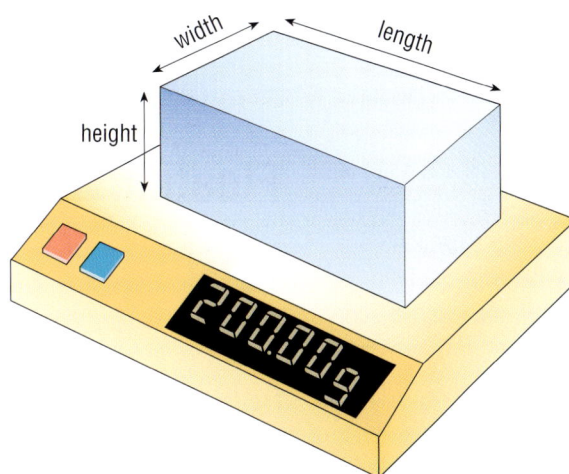

Figure 13.2 Block on a top-pan balance.

Measuring the density of a liquid

The density of a liquid is found in the following way.

- A measuring cylinder is put on a balance and its mass found (A).
- The liquid is poured into the measuring cylinder and its volume measured (V).
- The mass of the measuring cylinder and the liquid it contains is found (B).
- The mass of the liquid is found by subtracting A from B.
- The density of the liquid is calculated by dividing the mass of the liquid by its volume:

$$\frac{B - A}{V}$$

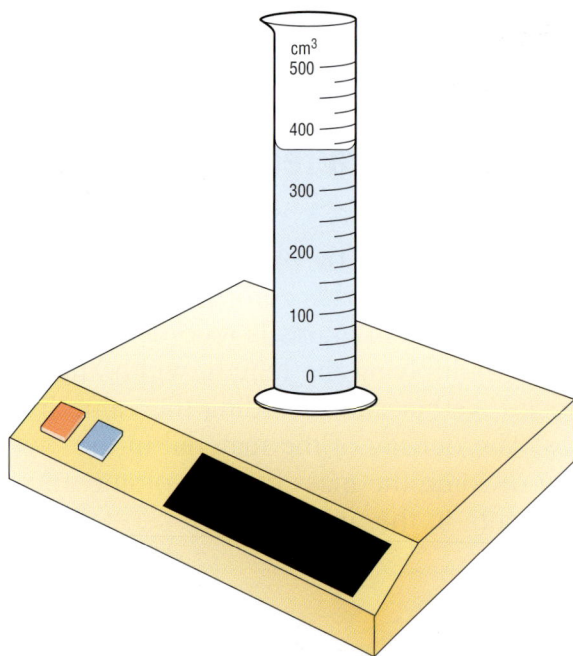

Figure 13.3 Mass of the measuring cylinder and the liquid it contains is measured.

Table 13.2 shows the densities of some liquids.

Table 13.2 The densities of some liquids.

Liquid	Density/kg/m^3
mercury	13 550
water at 4°C	1000
corn oil	900
turpentine	860
paraffin oil	800
methylated spirits	790

6 When paraffin oil and water are poured into a container they separate and the paraffin oil forms a layer on top of the water. When water and mercury are mixed the water forms a layer on top of the mercury.
   a) What can you conclude from these two observations?
   b) What do you predict would happen if water and corn oil were mixed together? (Refer to Table 13.2.)
7 What do you think would happen if the following solids were placed in water:
   a) expanded polystyrene,
   b) polythene,
   c) perspex?
   Explain your answers (refer to the values in Table 13.1).
8 What do you think would happen if the following solids were placed in mercury:
   a) steel, b) gold, c) lead?
   Explain your answers.

Floating and sinking

When a piece of wood is placed in water the wood floats. This is due to the difference in the densities of the wood and the water. From Tables 13.1 and 13.2 you can see that wood is less dense than water. When two substances, such as a solid and a liquid or a liquid and a liquid, are put together the less dense substance floats above the denser substance.

When full-fat milk is poured into a container, such as a bottle, the cream, which contains fat and is less dense than the more watery milk, rises to the top.

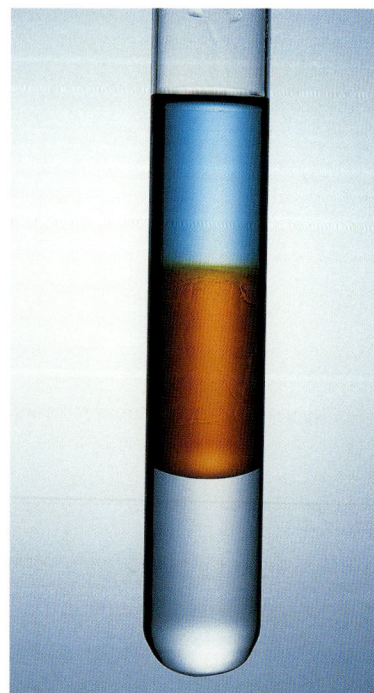

Figure 13.4 Liquids of different densities form layers when they are mixed.

9 Why do you think the temperature of water is shown when the value of its density is given?

10 Most people can just about float in water (Figure 13.5). What does this tell you about the density of the human body?

Figure 13.5

11 When salt is dissolved in water the solution that is produced has a greater density than pure water. An object that floats on pure water is shown in Figure 13.6. When it is placed in salt solution do you predict it will rise higher in the solution than it did in pure water or sink lower?

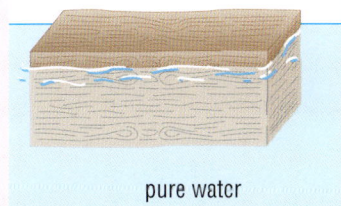

pure water

Figure 13.6

12 How is the process of finding the mass of a gas different from that of finding the mass of a liquid? Why is the difference necessary?

13 How can gas density be used to explain why hydrogen rises in air and carbon dioxide sinks?

Density of gases

Air is a mixture of gases. Its density can be found in the following way.

- The mass of a round-bottomed flask with its stopper, pipe and closed clip is found by placing it on a sensitive top-pan balance.

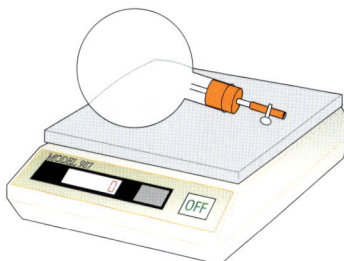

Figure 13.7 Measuring the mass of a flask on a top-pan balance.

- The flask is then attached to a vacuum pump and the air is removed from the flask and the clip is closed.
- The mass of the evacuated flask, stopper pipe and closed clip is found by placing it back on the balance. The mass of the air in the flask is found by subtracting the second reading from the first.
- The volume of the air removed is found by opening the clip under water so that water enters to replace the vacuum. The water is then poured into a measuring cylinder to find the volume.

Table 13.3 shows the densities of some gases.

Table 13.3 The densities of some gases.

Gas	Density/kg/m^3
hydrogen	0.089
air	1.29
oxygen	1.43
carbon dioxide	1.98

The density of a gas changes as its temperature and pressure change. The densities of gases are compared by measuring them at the same temperature and pressure. This is called the standard temperature and pressure (STP). The standard temperature is 0°C. The standard pressure of a gas is that pressure that will support 760 mm of mercury in a vertical tube.

When two gases meet the less dense gas rises above the denser gas.

Pressure on a surface

In earlier chapters we have examined forces acting at a point on an object. In this chapter we consider the effect of a force acting over an area.

When a force is exerted over an area we describe the effect in terms of pressure. Pressure can be defined by the equation:

$$\text{pressure} = \frac{\text{force}}{\text{area}}$$

The SI unit for pressure is N/m^2 but it can also be measured in N/cm^2.

An object resting on a surface exerts pressure on the surface because of the object's weight. Weight is the force produced by gravity acting on a solid, a liquid or a gas, pulling the material downwards towards the centre of the Earth. The weight acts on the mass of that material. For example, the weight of a solid cube acts on that cube (Figure 13.8).

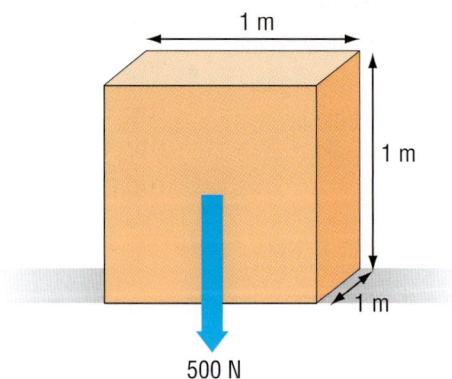

Figure 13.8   *The weight acting on a cube of material.*

The cube pushes down on the ground (or other surface that it rests on) with a force equal to its weight. The pressure that the cube exerts on the ground is found by using the equation above. For example, if the cube has a weight of 500 N and the area of its side is $1\ m^2$, the pressure it exerts on the ground is:

$$\text{pressure} = 500/1 = 500\ N/m^2$$

If the cube had a weight of 500 N and the area of its side was $2\ m^2$, the pressure it would exert on the ground is:

$$\text{pressure} = 500/2 = 250\ N/m^2$$

14 What is the pressure exerted on the ground by a cube which has a weight of 600 N and a side area of
  a) 1 m^2,
  b) 3 m^2?

15 What is the pressure exerted on the ground by an object which has a weight of 50 N and a surface area in contact with the ground of
  a) 1 cm^2,
  b) 10 cm^2,
  c) 25 cm^2?

16 a) What pressure does a block of weight 600 N and dimensions 1 m × 1 m × 3 m exert when it is
    i) laid on its side,
    ii) stood on one end?
  b) Why does it exert different pressures in different positions?

An object exerts a pressure on the ground according to the area of its surface that is in contact with the ground. For example, a block with dimensions 1 m × 1 m × 2 m and a weight of 200 N will exert a pressure of 200/1 = 200 N/m^2 when it is stood on one end (Figure 13.9a) but a pressure of only 200/2 = 100 N/m^2 when laid on its side (Figure 13.9b).

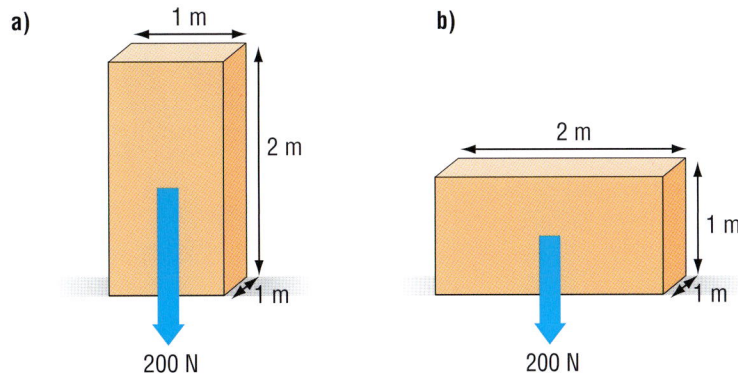

Figure 13.9 The weight acting on a block in two positions.

Your weight acting downwards causes you to exert a force on the ground through the soles of your shoes. If you lie down this force acts over all the areas of your body in contact with the ground. These areas together are larger than the areas of the soles of your shoes and you therefore push on the ground with less pressure when lying down than when you are standing up.

Figure 13.10 The force you exert downwards acts over a larger area when you lie down.

Reducing the pressure

When people wear skis the force due to their weight acts over a much larger area than the soles of a pair of shoes. This reduces the pressure on the soft surface of the snow and allows the skier to slide over it without sinking.

17 Drivers in Iceland, when going out on the snow, let their tyres down until they are very soft. The tyres spread out over the surface of the snow as they drive along. Why do you think the drivers do this?

Figure 13.11   Skis stop you sinking into the snow.

Increasing the pressure

Studs

Sports boots for soccer and hockey have studs on their soles. They reduce the area in contact between your feet and the ground. When you wear a pair of these boots your downward force acts over a smaller area than the soles of your feet and you press on the ground with increased pressure. Your feet sink into the turf on the pitch and grip the surface more firmly. This makes it easier to run about without slipping while you play the game.

Figure 13.12   The studs on this soccer boot help the player to grip the turf.

Pins and spikes

When you push a drawing pin into a board the force of your thumb is spread out over the head of the pin so the low pressure does not hurt you. The same force, however, acts at the tiny area of the pin point. The high pressure at the pin point forces the pin into the board.

18 A girl wearing trainers does not sink into the lawn as she walks across it but later when she is wearing high-heeled shoes she sinks into the turf. Why does this happen?

Sprinters use sports shoes which have spikes in their soles. The spike tips have a very small area in contact with the ground. The weight of the sprinter produces a downward force through this small area and the high pressure pushes the spikes into the hard track, so the sprinter's feet do not slip when running fast.

Figure 13.13   The spikes stop the sprinter from slipping on the track.

Knives

As we have seen, high pressure is made by having a large force act over a small area. The edge of a sharp knife blade has a very small area but the edge of a blunt knife blade is larger. If the same force is applied to each knife the sharp blade will exert greater pressure on the material it is cutting than the blunt knife blade and will cut more easily than the blunt blade.

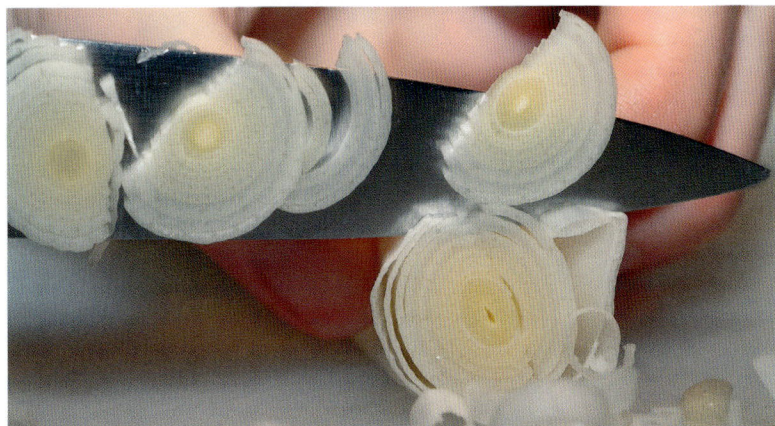

Figure 13.14   Knives cut well when they are sharp because of the high pressure under the blade.

Particles and pressure

Matter is made from particles. In solids the particles are held in position. In liquids the particles are free to slide over each other and in gases the particles are free to move away from each other.

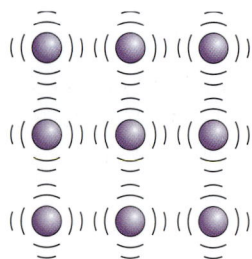

solid particles vibrate
to and fro

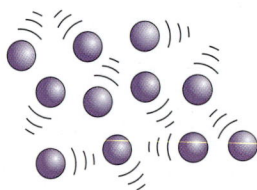

liquid particles have some
freedom and can move
around each other

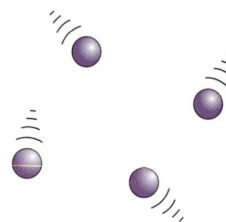

gas particles move freely
and at high speed

Figure 13.15   Arrangement of particles in a solid, a liquid and a gas.

Pressure in a liquid

In a solid object the pressure of the particles acts through the area in contact with the ground. In a liquid the pressure of the particles acts not only on the bottom of the container but on the sides too (Figure 13.16).

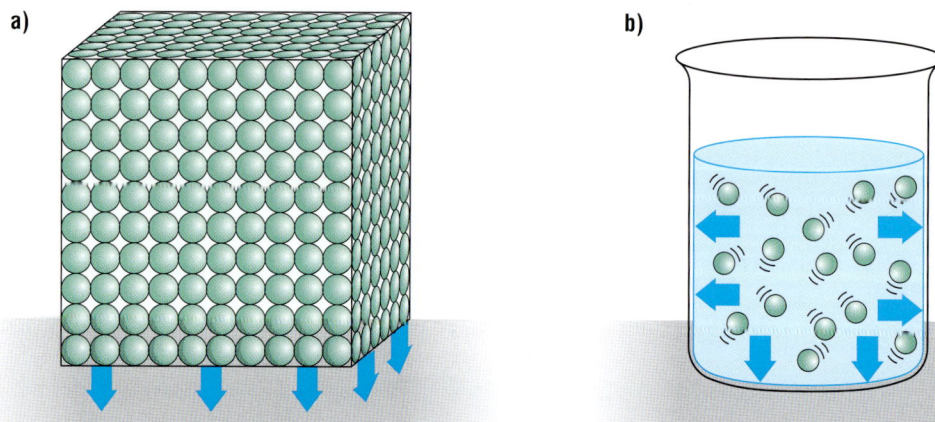

a)

b)

Figure 13.16   Pressure exerted by **a)** particles in a solid block and **b)** particles in a liquid.

Pressure and depth of a liquid

The change in pressure with depth in a liquid can be demonstrated by setting up a can as shown in Figure 13.17. When the clips are removed from the three rubber tubes, water flows out as shown. All three jets of

19 How does the path of the jet of water at the bottom of the can in Figure 13.17 change as the water level in the can falls? Why does it change?

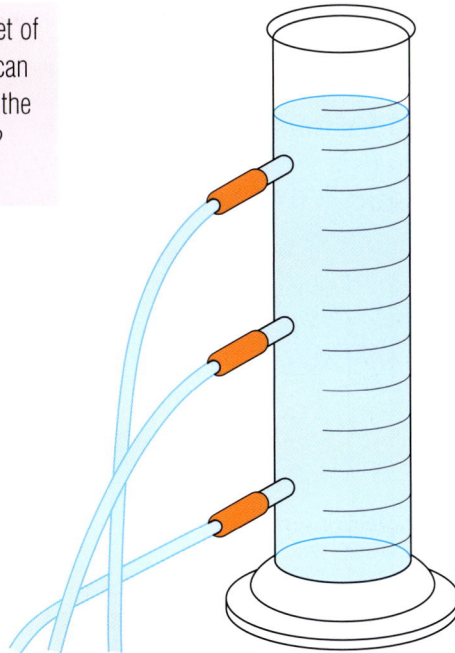

water leave the can horizontally but the force of gravity pulls them down. The water under the greatest pressure travels the furthest horizontally before it is pulled down. The water under the least pressure travels the shortest distance horizontally before it is pulled down.

Figure 13.17 Jets of water leaving a can at different depths.

A closer look at pressure and depth

The mass of a cubic centimetre of water is 1 g or 0.001 kg. The force of gravity (10 N/kg) means that this mass exerts a force downwards. The size of the force (equal to its weight) is calculated by:

$$0.001 \times 10 = 0.01 \text{ N}$$

This force acts on an area of 1 cm^2 (Figure 13.18a) so the pressure it exerts is:

$$\frac{0.01}{1} = 0.01 \text{ N/cm}^2$$

If a second cube of water is placed over the first, the pressure beneath the lower cube is increased to 0.02 N/cm^2 since the weight of water has doubled but the area it rests on has not (Figure 13.18b).

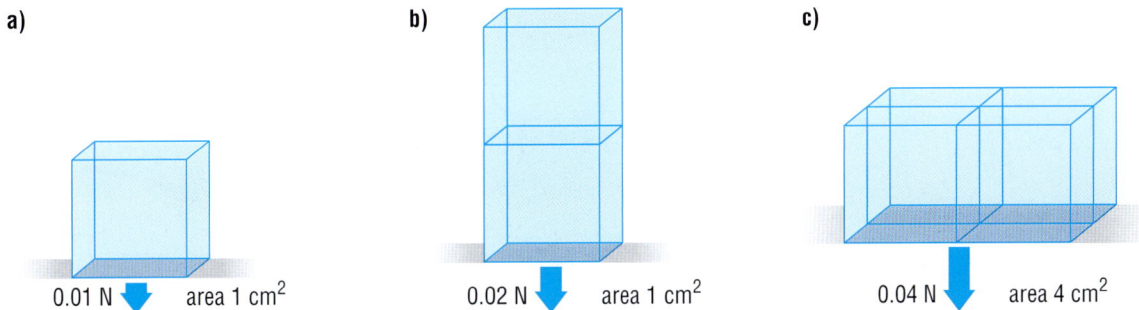

a)

0.01 N   area 1 cm^2

b)

0.02 N   area 1 cm^2

c)

0.04 N   area 4 cm^2

Figure 13.18 The pressure doubles when the depth of water doubles, but the pressure does not depend on the area of the column of water.

In fact, the pressure exerted by a liquid depends on the height of the column of liquid above its base, no matter what the area of the base of the column. Consider four cubes of water placed as in Figure 13.18c. A force of 0.04 N acts over an area of 4 cm^2 so the pressure is 0.01 N/cm^2, as in Figure 13.18a.

If water is placed in the two arms of a vessel as shown in Figure 13.19 and the partition between the arms is removed, the water moves down the left arm and up the right arm until the water in both arms is at the same level. When this happens both columns of water are exerting the same pressure on the bottom of the vessel.

You can see that although the arms are of different widths the water in them settles to the same level in each. The wider arm has more of the liquid in it but it also has a larger area. The water level indicator on some jug kettles uses this fact to allow you to see where the water level is inside.

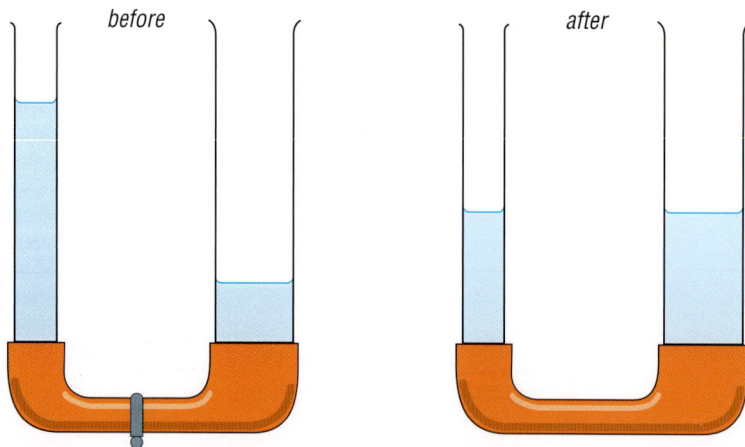

Figure 13.19   Water flows until the pressure in each column is the same.

20 Why does a dam need a wall shaped like that in Figure 13.21?

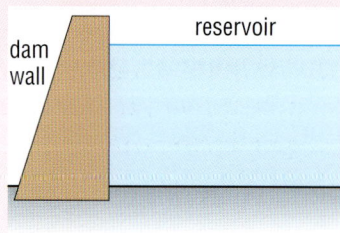

Figure 13.21   Cross-section of a dam wall.

21 Make a copy of Figure 13.22 and mark in the positions of the water levels in the different parts of the vessel B, C and D.

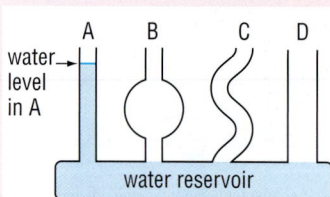

Figure 13.22   Pascal's vases (named after Blaise Pascal).

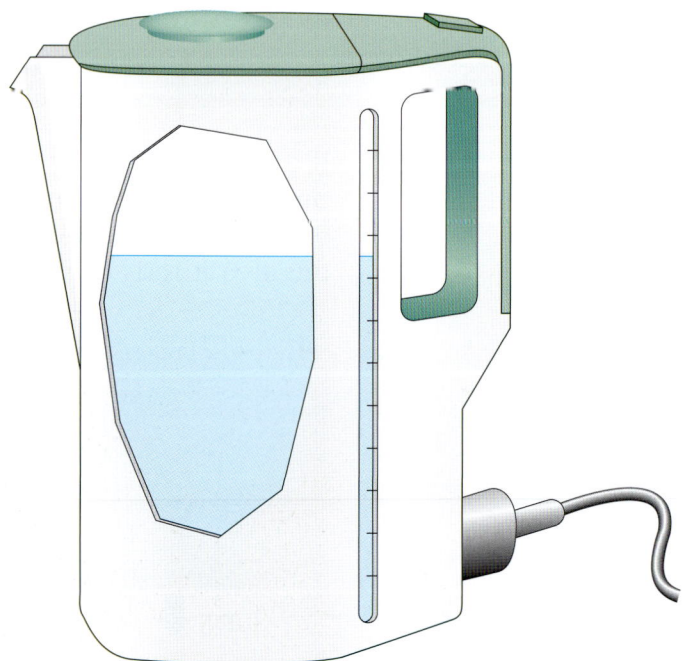

Figure 13.20   Inside a jug kettle.

Hydraulic equipment

If pressure is applied to the surface of a liquid in a container, the liquid is not squashed. It transmits the pressure so that pressure pushes on all parts of the container with equal strength.

In hydraulic equipment a liquid is used to transmit pressure from one place to another. The pressure is applied in one place and released in another. If the area where the pressure is applied is smaller than the area where the pressure is released, the strength of the force is increased as the following example shows.

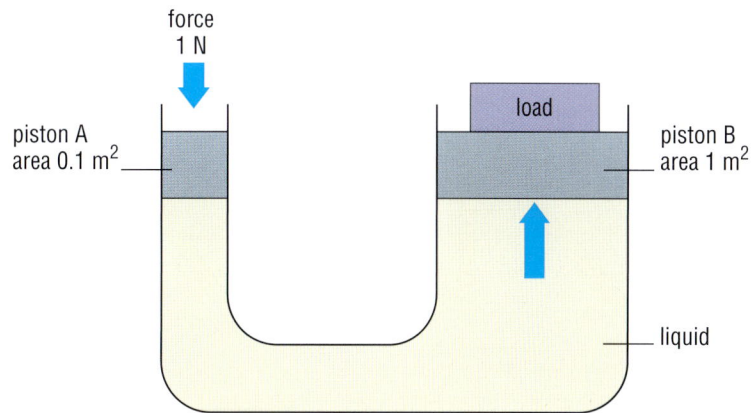

Figure 13.23   A simple hydraulic system.

22 Why are hydraulic systems known as 'force multipliers'?

A car may be raised with a small force by using a hydraulic jack. When a small force is applied to a small area of the liquid in the jack, a larger force is released across a larger area and acts to raise the car.

Figure 13.24   This car has been raised into the air for repairs by a hydraulic jack.

The brake system on a car is a hydraulic mechanism. The small force exerted by the driver's foot on the brake pedal is converted into a large force acting at the brake pads. This results in a large frictional force that makes it harder for the wheels to turn and so stops the car.

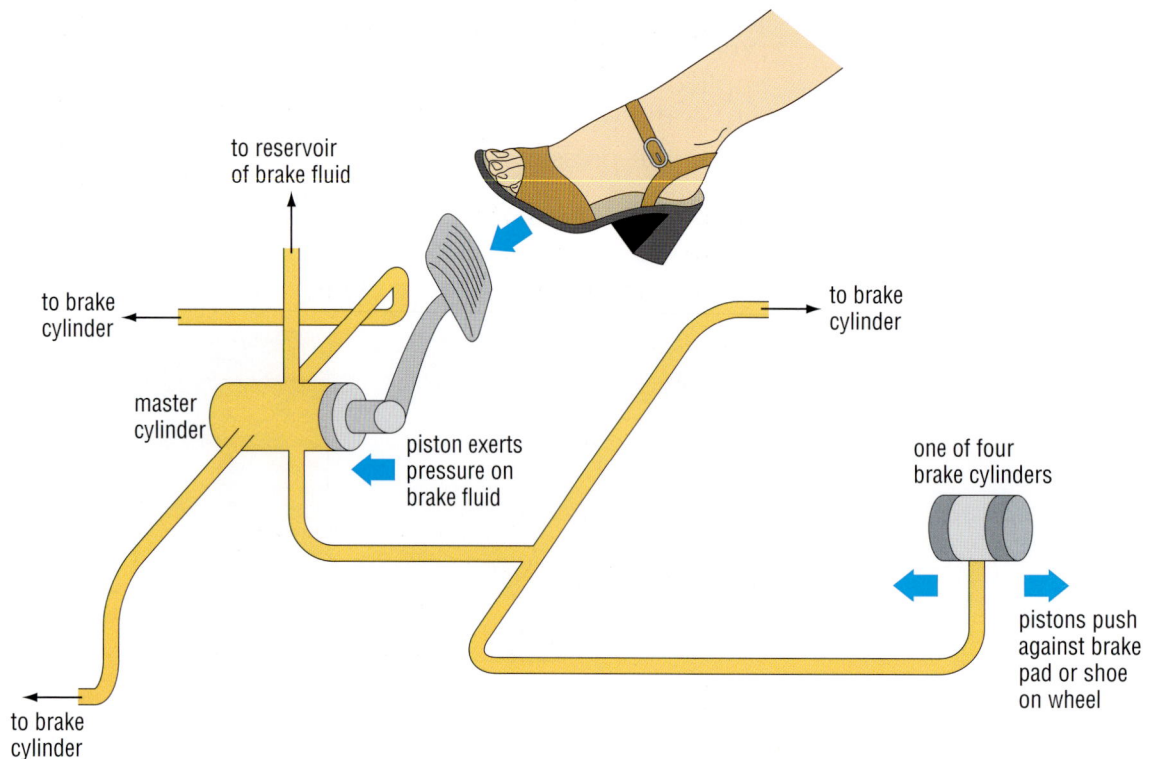

to reservoir
of brake fluid

to brake
cylinder

master
cylinder

piston exerts
pressure on
brake fluid

to brake
cylinder

one of four
brake cylinders

pistons push
against brake
pad or shoe
on wheel

to brake
cylinder

Figure 13.25   Hydraulic car brakes.

Pressure of the atmosphere

The atmosphere is a mixture of gases. The molecules from which the gases are made move around but are pulled down by the force of gravity exerted on them by the Earth. The atmosphere forms a layer of gases over the surface of the Earth which is about 1000 km high. This creates a pressure of about 100 000 N/m^2 – equivalent to a mass of 10 tonnes on 1 m^2 – although this gets less as you go up through the atmosphere.

You do not feel the weight of this layer of air above you pushing down because the pressure it exerts acts in all directions, as it does in a liquid. Thus, the air around you is pushing in all directions on all parts of your body. You are not squashed because the pressure of the blood flowing through your circulatory system (see *Checkpoint Biology* Chapter 8, page 114) is strong enough to balance atmospheric pressure.

The atmosphere does not crush ordinary objects around us. For example, the pressure of the air pushing down on a table top is balanced by the pressure of the air underneath the table pushing upwards on the table top.

Ear popping

The middle part of the ear (Figure 13.26) is normally filled with air at the same pressure as the air outside the body. The air pressure can adjust because when you swallow, the Eustachian tubes in your throat open and air freely enters or leaves the middle ear. For example, if the air pressure is greater outside the body and in the mouth, when you swallow more air will enter the middle ear to raise the air pressure there.

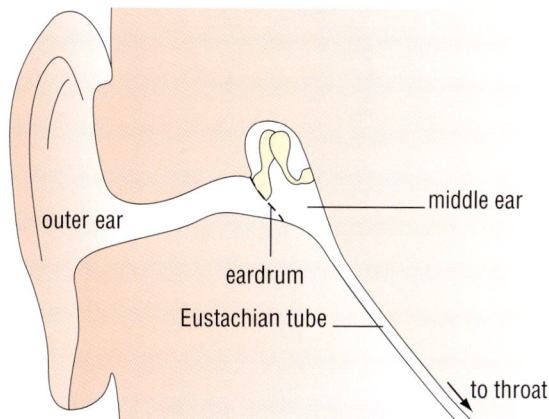

Figure 13.26   The ear and throat.

23 What happens if the air pressure in the throat and outside the body is less than the air pressure in your middle ear when you swallow?

24 If you ride quickly down a hill on a bicycle your eardrums are pushed in before they pop back. Why is this?

If you travel in a car which quickly climbs a steep hill, your ears sometimes 'pop'. This is because you are rising rapidly into the atmosphere where the pressure is lower. The popping sensation is caused by the air pressure being lower in the throat and outside the body than in the middle ear. The difference in pressure causes the eardrum to push outwards. When you swallow the air pressure in your middle ear reaches the same pressure as the air in your throat and outside, and the eardrum moves quickly back – or 'pops' – into place.

How a sucker sticks

When an arrow with a sucker on the end hits a target the arrow stays in place due to air pressure. As the elastic sucker hits the flat surface it deforms and pushes out some of the air from beneath the cup. The pressure of the remaining air in the cup is less than that of the air pressure outside the cup. The higher pressure of the air outside the cup holds the sucker in place (Figure 13.27).

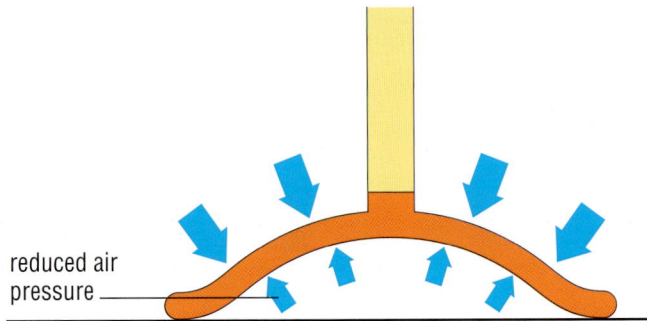

Figure 13.27 Side view through a sucker.

reduced air pressure

Crushing a can

The strength of the air pressure in the atmosphere can be demonstrated by taking the air out of a can. This can be done in two ways.

Using steam

The can has a small quantity of water poured into it and is heated from below. As the water turns to steam it rises and pushes the air out of the top of the can. If the heat source is removed and the top of the can immediately closed, the remaining steam and water vapour in the can will condense, leaving only a small quantity of air in the can. This air has a much lower pressure than the air pressure outside the can and the higher pressure crushes the can.

Using a vacuum pump

A vacuum pump can reduce the pressure in containers. If one is used to remove air from a can, the can collapses due to the greater pressure of the air on the outside (Figure 13.28).

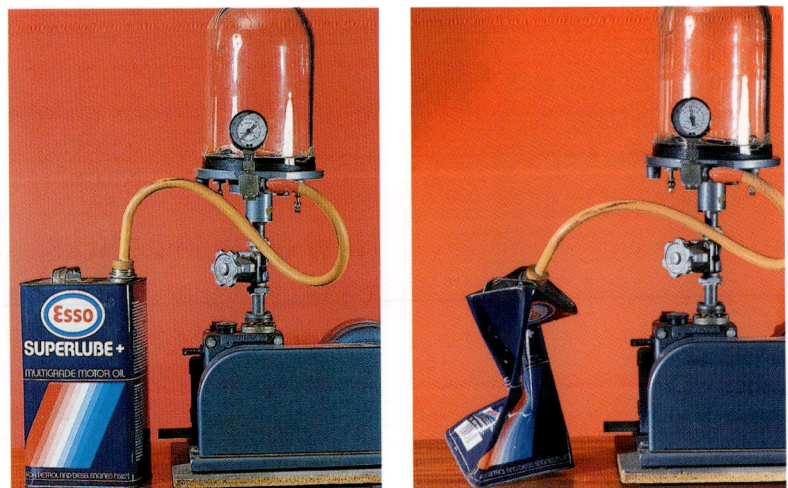

Figure 13.28 Removing air from this empty oil can has made it collapse.

A scientific showman

Otto von Guericke (1602–1686) was the mayor of Magdeburg in Germany for 35 years. He was also interested in science (see page 124). He was keen to discover if a vacuum could really exist and made an air pump to test his ideas. He used his pump to draw air out of a variety of vessels. When he tried barrels he found they collapsed. He eventually found that hollow copper hemispheres joined together to make a globe were much stronger.

Aristotle had believed that if a vacuum could exist then sound would not be able to pass through it. When von Guericke put a bell in one of his vessels and removed the air he discovered that a ringing bell could not be heard (see page 83).

Von Guericke greatly enjoyed demonstrating his discoveries to large numbers of people and in one demonstration had two teams of eight horses pull on the evacuated hemispheres without separating them. This spectacular demonstration helped people to realise the strength of the air pressure pushing on the hemispheres.

1 Why did von Guericke believe he had made a vacuum?
2 Why did the Magdeburg hemispheres not come apart when the teams of horses pulled on them?
3 How did von Guericke help people to become interested in science?

For discussion

How could an investigation you have done in your physics course be made into an informative and entertaining demonstration?

How successful are television producers and presenters at making science programmes entertaining and informative?

Figure A Otto von Guericke's demonstration with 'Magdeburg' hemispheres.

Aerosols

An aerosol spray can contains a gas which is at a higher pressure than air pressure. It is held in the can by a valve in the nozzle (Figure 13.29). When the nozzle is pressed down a spring is squashed and the nozzle opening enters the inside of the can, effectively opening the valve. The higher pressure of the gas in the can pushes on the liquid in the can and it rushes up the tube and through the jet where it forms a fine spray of liquid droplets (an 'aerosol'). When the nozzle is released the spring is no longer squashed and pushes the nozzle

25 How many uses of aerosol cans in the home can you think of?

upwards. This removes the nozzle opening from inside the can, effectively closing the valve, and stops the flow of spray.

Aerosol cans used to contain a gas made from chlorofluorocarbons (CFCs). These chemicals are now known to damage the ozone layer. In many countries they have now been replaced with gases such as refinery gases which do not damage the ozone layer.

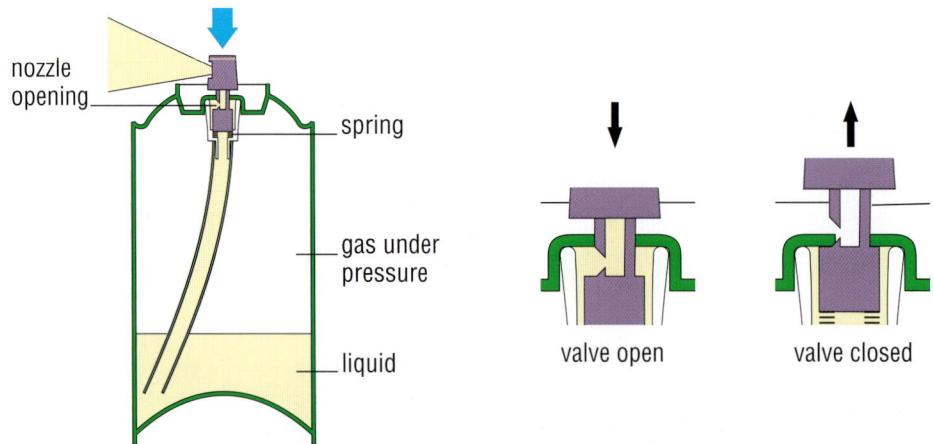

nozzle opening

spring

gas under pressure

liquid

valve open

valve closed

Figure 13.29 Inside an aerosol can.

Hovercraft

A hovercraft uses the pressure of air to raise it from the ground. It does this by drawing air from above with powerful fans. There is a skirt around the edge of the hovercraft which prevents the air from escaping quickly and the air pressure beneath the hovercraft increases. The upward pressure of the air trapped beneath the hovercraft lifts the hovercraft off the ground. The fans continue to spin to replace air that is lost from the edges of the skirt.

26 What are the advantages of using a hovercraft as a means of transport?

The cushion of air beneath the hovercraft reduces friction between it and the ground. The cushion of air is also maintained when the hovercraft moves over water. The forward or backward thrust on the hovercraft is provided by propellers in the air above the hovercraft.

Figure 13.30 A hovercraft.

◆ SUMMARY ◆

◆ The density of a substance is a measure of the amount of matter that is present in a certain volume of it (*see page 138*).

◆ The density of solids, liquids and gases can be measured in different ways (*see pages 139–141*).

◆ Pressure acts when a force acts over an area of surface (*see page 142*).

◆ When a solid object exerts a pressure on the surface below it, the smaller the area of contact, the greater the pressure (*see page 144*).

◆ Pressure in a liquid acts in all directions and increases with the depth of the liquid (*see page 146*).

◆ In hydraulic systems pressure is transmitted through a liquid (*see page 149*).

◆ The atmosphere exerts a pressure (*see page 150*).

◆ Air pressure is made use of in various devices such as pumps, suckers, aerosols and hovercraft (*see pages 151–154*).

End of chapter question

1 How could you explain the following using the model of air made up of particles which move freely?

a) How air pushes on a surface.

b) Why the pressure in an inflated tyre is higher than the air pressure outside.

c) Why a sucker stays in place on a flat surface.

The section on pressure on page 58 of *Checkpoint Chemistry* may help you answer this question.

14 Electric current

The voltage of a cell

1 Compare how an ammeter and a voltmeter are connected into a circuit by looking at Figure 11.13 on page 129 and Figure 14.2 on this page.

The ability of the cell to drive a current is measured by its voltage. This is indicated by a figure on the side of the cell with the letter V after it. The volt, symbol V, is the unit used to measure the difference in electrostatic potential energy, usually just referred to as 'potential', between two points. The voltage written on the side of the cell refers to the difference in potential between its positive and negative terminals. It is a measure of the electrical energy that the cell can give to the electrons in a circuit.

When cells are arranged in series with the positive terminal of one cell connected to the negative terminal of the next cell, the current-driving ability of the combined battery of cells can be calculated by adding their voltages. For example, two 1.5 V cells in series produce a voltage of 3 V. The two cells together give the electrons in the circuit twice as much electrical energy as each one would provide separately.

Figure 14.1 The voltage is clearly displayed on the packaging of cells and batteries.

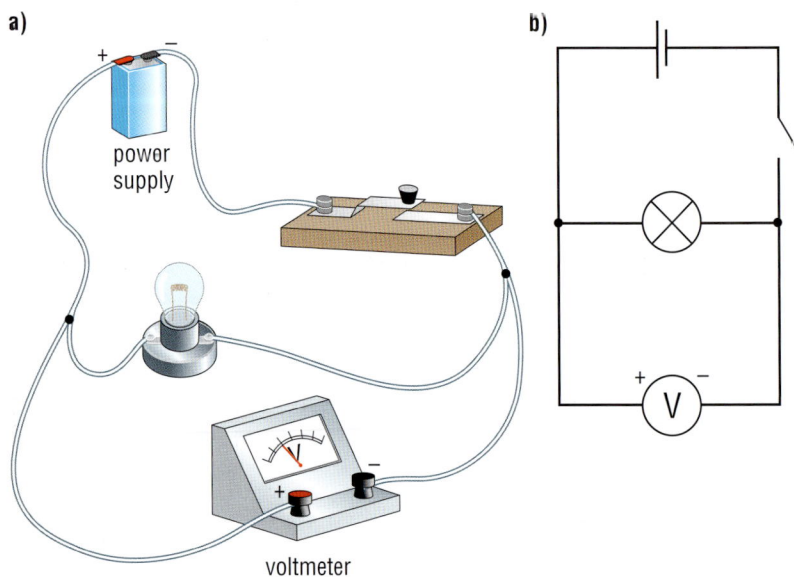

Figure 14.2 **a)** A voltmeter connected in a circuit and **b)** the circuit diagram showing its symbol.

Measuring voltage

The potential difference or voltage between two points in a circuit is measured by a voltmeter.

The voltmeter is connected into a circuit with its positive or red terminal connected to a wire that leads towards the positive terminal of the cell, battery or power pack. The negative or black terminal must be connected to a wire that

2 A wire carrying a current of electricity can be described as being similar to a stream carrying a current of water. In what ways are the wire and the stream similar?

3 Predict the brightness of the lamps in the circuits in Figure 14.4 compared with that of a single lamp in a circuit with one cell. Use one of the following words in each case:
very dim, dimmer, the same, brighter, very bright.
(All the lamps are identical and all the cells have the same voltage.)

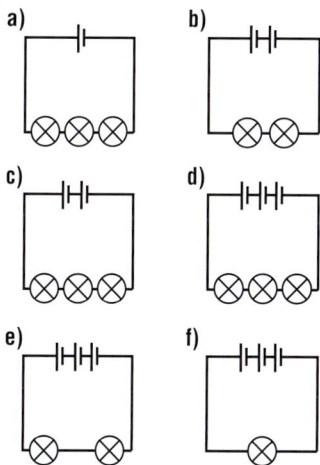

Figure 14.4

4 Compare the circuit in Figure 14.5 with the one in Figure 14.4b. Do you think the lamps will glow with the same brightness? Explain your answer.

Figure 14.5

leads towards the negative terminal of the source of the current. However, unlike the connection of an ammeter, the wires are attached to either side of the part of the circuit being tested – it is arranged in parallel with this part of the circuit. Voltmeters generally have a very high resistance, so when connected in parallel they take little current and do not affect the rest of the circuit.

Resistance

The material through which a current flows offers some resistance to the moving electrons. A material with a high resistance only allows a small current to pass through it when a certain voltage is applied. A material with a low resistance allows a larger current to pass through it for the same applied voltage.

The wires connecting the components in a circuit have a low resistance while the wires in the filaments of the lamps have a high resistance. When the lamps are connected in series their resistances combine in the same way as the voltages of the cells in series – they add. They therefore offer a greater resistance to the current than each lamp would separately.

Lamps and current size

The size of the current flowing though a circuit can be estimated by looking at the brightness of the lamps in the circuit. A lamp shines with normal brightness when it is connected to one cell as shown in Figure 14.3a. The lamp shines more brightly than normal when it is in a circuit with two cells (Figure 14.3b) and shines less brightly when it is in a circuit with one cell and another lamp as shown in Figure 14.3c.

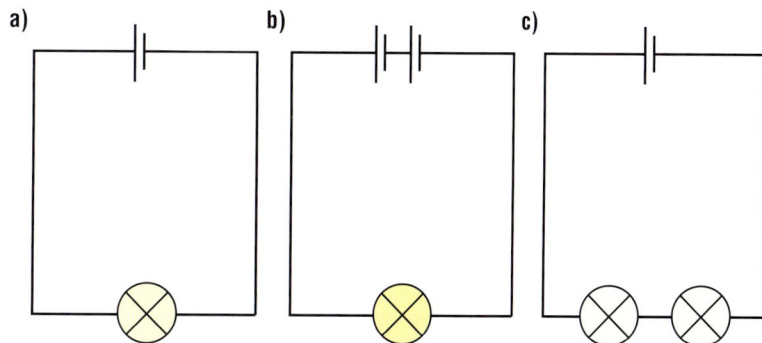

Figure 14.3 Three arrangements of cells and lamps in series.

5 How do you think the brightness of two lamps arranged in parallel compares with the brightness of two lamps arranged in series (both arrangements having one cell)?

6 If current flows through two lamps arranged
   a) in series,
   b) in parallel,
and the filament of one lamp breaks, what happens to the other lamp? Explain your answers.

Parallel circuits

Lamps can be arranged in a circuit 'side by side' rather than end to end. This kind of circuit is called a parallel circuit (Figure 14.6). The resistances of the lamps do not combine to oppose the flow of current in the same way as they do in a series circuit. Each lamp receives the same flow of electrons as it would if it were on its own in a circuit with the cell.

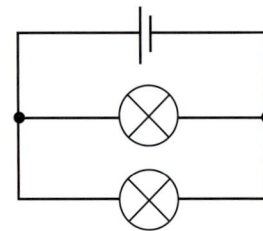

Figure 14.6   Two lamps in parallel.

The ammeter in parallel circuits

In a parallel circuit like the one shown in Figure 14.7 the ammeter reading varies in the following way. At points A and D the readings are the same. At points B and C, if the lamps are identical the readings will be the same and exactly half the readings at A and D. This means that if the readings at A and D are 4.0 amps the readings at B and C will be 2.0 amps.

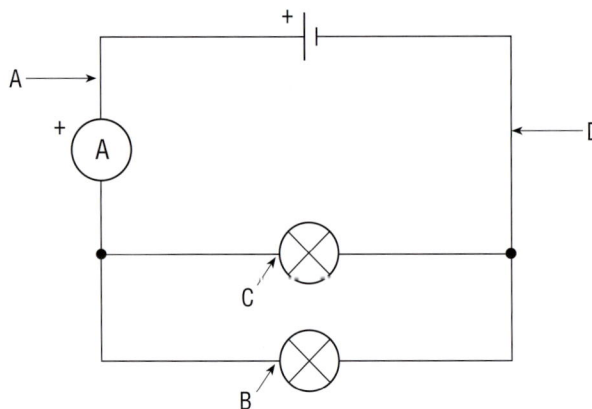

Figure 14.7   Using an ammeter to measure the current at different points in a parallel circuit.

The total of the readings of the ammeters at B and C is always the same as the reading at A and D even if the bulbs are not identical. For example if the reading at A and D is 4.0 amps and the reading at B is 3 amps then the reading at C will be 1 amp.

Light-dependent resistor

A light-dependent resistor (LDR) is made from two pieces of metal which are joined together by a semiconductor. A semiconductor is a material that has just a few electrons which can move freely.

For discussion

Which pieces of equipment in your home do you think have resistors which control the flow of electricity? You may like to look back at pages 40–41 to help you answer.

When the LDR receives light energy, more electrons are released in the semiconductor to move freely through it, and the resistance of the LDR becomes lower. When the amount of light shining on it is reduced, fewer electrons can flow and the resistance increases.

7 Some bedside clocks have a display which glows dimly when the room is dark yet shines brightly if the room is light. How could an LDR be responsible? How may the LDR make it easier for you to get to sleep?

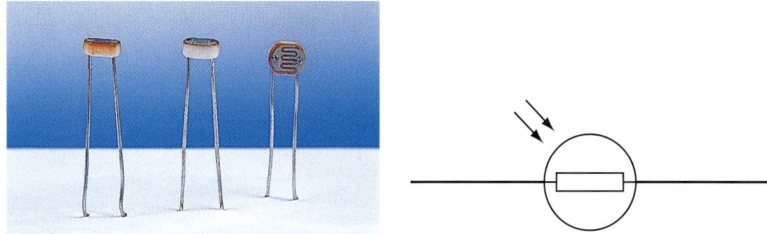

Figure 14.8 Three LDRs and the symbol for an LDR.

Diodes

A current can only flow through a diode in one direction. They are used to control the direction of the flow of a current through complicated circuits, such as those used in a radio, which have components in series and parallel.

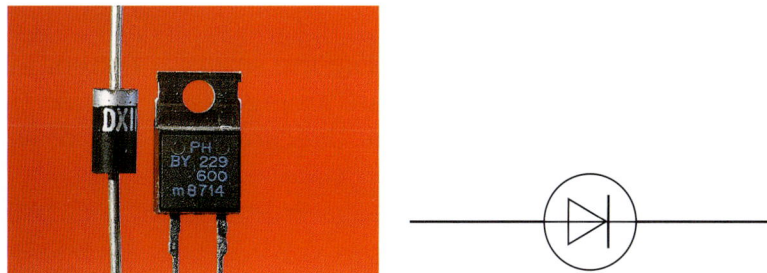

Figure 14.9 Two diodes and the symbol for a diode.

8 Compare the actions of a resistor and a diode.

Diodes have a band marked at one end. When the diode is connected into a circuit the end with the band on must be connected to a wire coming from the negative terminal of the cell or battery for the current to flow. When a diode symbol is drawn in a circuit diagram the symbol should be drawn with the straight line facing the negative terminal of the source of the current.

Light-emitting diodes

A light emitting diode is often referred to as an LED. In simple circuits we often use a lamp to show that a current is flowing. In electronic circuits an LED performs the same task more efficiently. An LED is a semiconductor diode, allowing a current to flow in only one direction through it, and it produces light. An LED can emit red, yellow or green light. The colour emitted depends on the semiconductor materials used to make the LED.

9 How is an LED
  a) similar to, and
  b) different from a lamp?

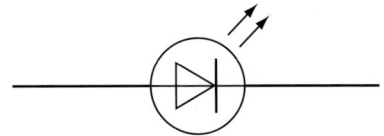

Figure 14.10 Three LEDs and the symbol for an LED.

◆ SUMMARY ◆

◆ The volt is the unit used to measure the potential difference between two points (*see page 156*).

◆ A voltmeter is used to measure the voltage between two points in a circuit (*see page 156*).

◆ The material through which a current flows offers some resistance to the moving electrons (*see page 157*).

◆ Lamps can be used to estimate the size of a current in a circuit (*see page 157*).

◆ In a parallel circuit, lamps are arranged side by side (*see page 158*).

◆ A diode is use to control the direction of current flow through a circuit (*see page 159*).

End of chapter question

1 A circuit contains a cell, an LDR, an LED and a switch. When the switch is closed and the circuit is left in daylight the LED glows, but when the closed circuit is left in the dark the LED no longer glows. Explain what is happening in the circuit in both the light and the dark.

15 *Energy crisis*

We are moving towards a time when there will be an energy crisis. A crisis is a time of danger or of great difficulty. We are in danger of not having enough energy for our needs and this will create great difficulty.

Energy pathways on Earth

The Sun sends out a huge amount of energy into space, in all directions, every second. Only a tiny fraction of this energy reaches the Earth. Even so, this amount of energy from the Sun is large in our terms on Earth. It would take 173 million large power stations to produce the same amount of energy that we receive from the Sun each second. However, not all of the Sun's energy that reaches the Earth can be useful to us.

Figure 15.1 on the next page shows a Sankey diagram about energy pathways on Earth. It features energy from the Sun, geothermal energy and energy from the tides. It does not include nuclear energy which is described on pages 168–169.

The water cycle

Nearly a quarter of the Sun's energy drives the water cycle on Earth. It heats the surface of the water in the oceans and seas causing large amounts of water to evaporate. The water vapour passes into the air and eventually condenses on dust. The tiny droplets of condensed water form clouds. Winds push the clouds along and when they reach cool regions such as high mountains, larger droplets form and rain is released. The rainwater forms rivers, which lead back to the seas.

Winds and waves

When the Sun's rays heat the ground some of the heat is passed into the air. The warm air rises and is replaced by cooler air. This movement of the air generates winds. Some of them blow on the surfaces of seas and cause waves to form. The movement of the air and the up and down movement of the waves can be used to generate electrical energy (see pages 31 and 171).

For discussion

Think about this before moving on to read the rest of the chapter. What do people think will happen when the energy crisis comes? How do they think they could try and avoid the crisis happening?

1 How much of the Sun's energy is reflected back into space or absorbed by the atmosphere?
2 Make a drawing of the water cycle from the information in the paragraph on the right.
3 The water at the head of a river is much higher than the water at sea level. What kind of energy does it possess? (Look at page 33 in Chapter 1 to help you.)
4 How may the energy possessed by moving water be used to provide us with energy we can use? (Look at page 98 in Chapter 9 to help you answer this.)

For discussion
What are the sources of the energy that reaches the Earth's surface? If each one in turn was no longer available, could we survive?

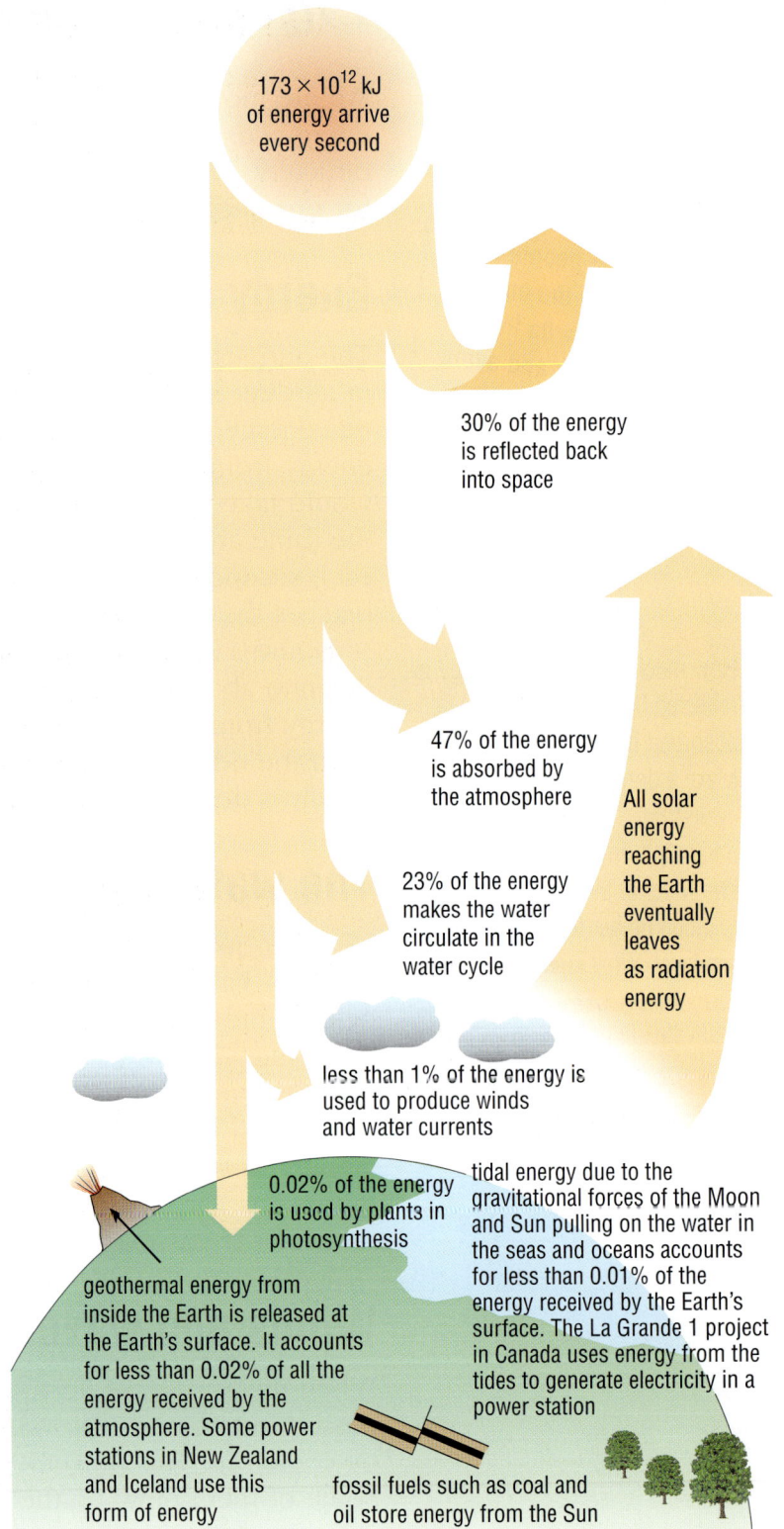

173×10^{12} kJ of energy arrive every second

30% of the energy is reflected back into space

47% of the energy is absorbed by the atmosphere

All solar energy reaching the Earth eventually leaves as radiation energy

23% of the energy makes the water circulate in the water cycle

less than 1% of the energy is used to produce winds and water currents

0.02% of the energy is used by plants in photosynthesis

tidal energy due to the gravitational forces of the Moon and Sun pulling on the water in the seas and oceans accounts for less than 0.01% of the energy received by the Earth's surface. The La Grande 1 project in Canada uses energy from the tides to generate electricity in a power station

geothermal energy from inside the Earth is released at the Earth's surface. It accounts for less than 0.02% of all the energy received by the atmosphere. Some power stations in New Zealand and Iceland use this form of energy

fossil fuels such as coal and oil store energy from the Sun

Figure 15.1 Sankey diagram showing the flow of energy to and from the Earth's surface.

Tidal and geothermal energy

There are two energy paths which do not begin with the Sun. One energy path begins with the movement energy in the tides as they rise and fall due to the pull of gravity of the Moon and Sun on the waters of the seas and oceans. The second energy path begins with the heat energy released by radioactive materials (see page 168) inside the Earth.

Living things and fossil fuels

Only a very small amount of the Sun's energy is trapped by plants and used to make food. The food is used by plants and animals. When plants and animals die, their bodies decompose and the energy that they possessed passes into the decomposing organisms, such as bacteria, and eventually passes out into the air and space as heat. In the past there were regions in the world where this did not happen. The energy in the bodies of ancient trees and marine organisms was trapped underground in their bodies. These organisms did not decay and fossils fuels (coal, oil and gas) were formed (see page 35).

5 What is the name of the process by which plants make food from the energy in sunlight?

Figure 15.2 The conditions under which coal forms produce different types of coal as these photos show – anthracite (left) and vitranite (right).

Figure 15.3 These bottles contain samples of oil from different parts of the world.

From one crisis to another

The first fuel that humans used was wood. There was plenty of it around and the trees grew back at the rate at which people used them. In time, in some countries, the numbers of people increased and the demand for fuel increased as well. The trees could not provide enough energy to meet the needs of the people and an energy crisis loomed. It was averted by the discovery of fossil fuels – coal, oil and gas. They are the fuels on which much of the world depends today. The problem with a fossil fuel is that it takes millions of years to develop. It cannot form at a rate which matches the rate at which we use it up, so eventually it will all be gone.

The problems with fossil fuels

It is quite easy for people to collect wood for fuel. They can just pick up dead branches or chop down a tree. The fuel has then only to be transported, possibly a short distance, to the place where it will be used. When people use fossil fuels there are more problems to overcome.

Extracting coal

Coal forms in layers below the ground. The layers are called seams. If the coal seam is near the surface of the ground, the soil and rocks above it are removed and an open-cast mine is formed. The coal seam is removed and the rocks and soil are replaced.

If the coal seams are deep, shafts are sunk to meet them and then tunnels are dug into them. Coal is dug out of the seam at a wall called the coal face. It is transported back down the tunnel on trucks or conveyer belts and brought to the surface in lifts. Miners travel in and out of the mines on lifts too. Two dangers in the mine are firedamp and flooding. Methane gas escapes from the coal as it is being mined and mixes

Figure 15.4 Open-cast mining.

6 Look at Figure 15.5 and imagine that you are to about to set up a mine to extract coal from deep underground. What will you have to spend money on before you can begin extracting the coal?

with the air to form firedamp. If this mixture is lit it explodes. The build up of firedamp in a mine is avoided by sinking ventilation shafts and driving a current of air through the shafts and tunnels. Flooding is prevented by building sumps (pits into which water drains) at the bottom of shafts and pumping the water that collects there to the surface.

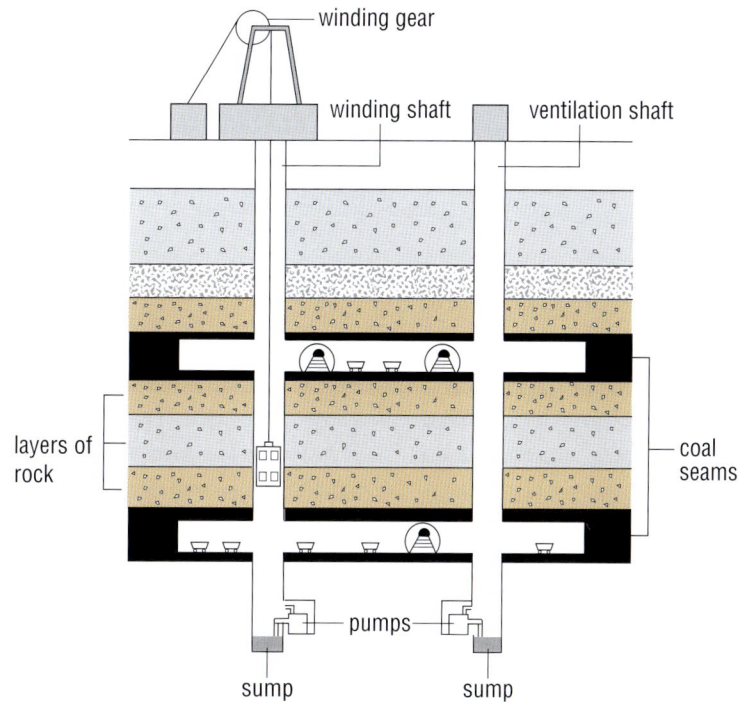

Figure 15.5   Mining deep coal seams.

7 Imagine that you have been given the task of extracting and selling oil from an oil well. You have to set up the oil well and transport the oil to the buyers. Look at the two kinds of oil well in Figure 15.6 a and b. Which kind do you think will be cheaper to set up? Explain your answer.

Extracting oil and gas

Oil and gas collects underground in some places where a porous rock is sandwiched between two layers of non-porous rock. A narrow shaft is drilled through the upper layer of non-porous rock to reach the layer of porous rock. The gas and oil pass up the shaft and are carried away along pipes to be stored and transported. Some gas and oil is present under the land in certain countries while in other places it is present under the seabed. Extracting oil from under the sea requires the

Figure 15.6a   An oil well on land.

Figure 15.6b An oil well under the sea.

8 Your oil well is running dry and you get the chance to set up an oil well 100 kilometres further away over some rugged hills where there are no roads. What extra expenses would you have to pay to extract this oil? How would the price of oil affect your decision to extract the oil?

building of a platform on which a drill can be set up. The platform also provides a home for the people working on drilling and extracting the fuels.

The price of fuel

The price of fuel is related to the cost of extracting it. It is cheaper to extract a fuel like coal when it is near the surface as the shafts are short and less energy is used on lifting the fuel to the surface. When the coal seams near the surface are used up, deeper seams have to be mined. This means making deeper shafts and using more energy lifting the fuel to the surface. This makes the fuel more expensive. In time it can become too expensive compared to other fuels and the mine is closed.

Mines and wells are set up close to places where the fuels are used or where they can be transported cheaply to the people who need them. As these mines and wells are used up other places where the fuels can be found may be used. These places may be more difficult to reach and make the cost of setting up mines and wells and transporting the fuels more expensive.

Fossil fuels and the environment

When fossil fuels burn they produce carbon dioxide. As we use large amounts of fossil fuels, large amounts of carbon dioxide are produced. This gas mixes with the gases in the air. The presence of carbon dioxide in the air makes the atmosphere behave like the glass in a greenhouse (see page 134). As extra carbon dioxide is

Figure 15.7 A rise in temperature, possibly due to global warming, has killed this tree.

For discussion

There have been periods in the Earth's history when conditions were warmer than today. At those times fossil fuels were not being burnt. Some scientists believe that global warming now is not due to fossil fuels. Others believe that fossil fuels are the major cause. Research the Internet for different views. What is the view that is most common today? Should people cut down on the use of fossil fuels?

produced this effect is increased. This causes the temperature of the atmosphere to rise and results in an increase in global warming.

Fossil fuels contain an element called sulphur. When the fuels are burnt the sulphur combines with oxygen to form sulphur dioxide. This reacts with oxygen and water vapour in the air and forms sulphuric acid. It may fall to the ground as acid rain. When acid rain passes into the soil it removes some of the minerals from the soil that plants need for healthy growth. The plants' growth slows and they become stunted or even die. When the acid rain reaches rivers and lakes it causes the acidity of the water to increase. Many aquatic organisms cannot adapt to the increase in acidity and die. In towns, the ornamental stonework on buildings may be destroyed by acid rain as shown in Figure 15.8.

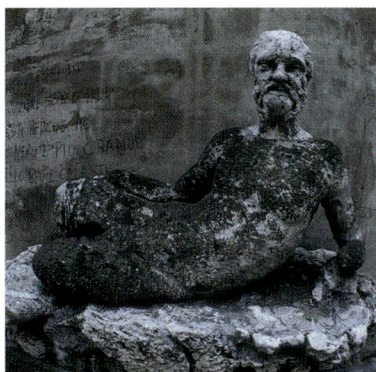

Figure 15.8 Acid rain has damaged this stone statue.

Non-renewable fuels

Fossil fuels are sources of non-renewable energy. They cannot be replaced because the rate at which they form is much less than the rate at which they are being used up. Radioactive materials are also a non-renewable energy source.

Radioactive materials

Almost all the elements which form the matter from which the Earth is made have been made by nuclear fusion in stars in the distant past. Some elements are unstable and 'decay' in nuclear reactions to form more stable elements. These unstable elements are called radioactive elements. When they undergo radioactive decay they release particles and energy.

Using nuclear fission

Albert Einstein (1879–1955), a German-born self-taught physicist, began by studying the work of other scientists. Among other enlightened theories, he produced an equation which linked mass and energy. The equation is:

$$E = mc^2$$

where E is energy, m is mass and c is the velocity of light.

This equation showed that mass is a store of energy which can be released when matter is destroyed. This idea successfully explained the phenomenon of radioactive materials by showing that they produced energy in the form of radiation because part of their mass was destroyed.

Otto Hahn (1879–1968), a German chemist, and Lise Meitner (1878–1968), an Austrian physicist, studied how radioactive materials decayed and in 1939 Meitner described how uranium atoms broke in half.

Figure A   Albert Einstein.   **Figure B**   Lise Meitner.

Leo Szilard (1898–1964), a Hungarian-born physicist who studied atomic nuclei, had an idea of how the break-up of one atom may cause the break-up of surrounding atoms and lead to a chain reaction. When Szilard heard of Hahn's and Meitner's work he realised that uranium could be investigated to see if a chain reaction would be set up. If such a reaction could take place then a large amount of energy could be released quickly.

Szilard was drawing up his plans and predictions in the United States during the Second World War. He realised that the chain reaction could be used to release energy for a bomb. He and other scientists, including Einstein, informed the United States president of the explosive power of uranium. This led to the development of nuclear bombs. Szilard and many scientists believed the bombs were too devastating to be used like other bombs and should only be demonstrated in an uninhabited part of the world, to show the enemy the power that was now set against them. However, politicians in charge of the armed forces and some scientists believed it right to use the nuclear bombs like other bombs, and two were dropped on Japan with catastrophic results.

1 Why do radioactive materials release energy?
2 What was Szilard's idea and how was the work of Hahn and Meitner useful to him?
3 Why do you think the scientists first considered making nuclear bombs instead of reactors for power stations?

Figure C   Devastation caused by a nuclear explosion in Hiroshima, Japan.

After the War scientists began investigating ways of using nuclear energy for peaceful purposes. This meant that the energy released in the chain reaction had to be released more slowly than in a bomb.

4 What could be the consequences of an explosion at a nuclear reactor?

5 What are the advantages and disadvantages of using nuclear fuel for generating electricity?

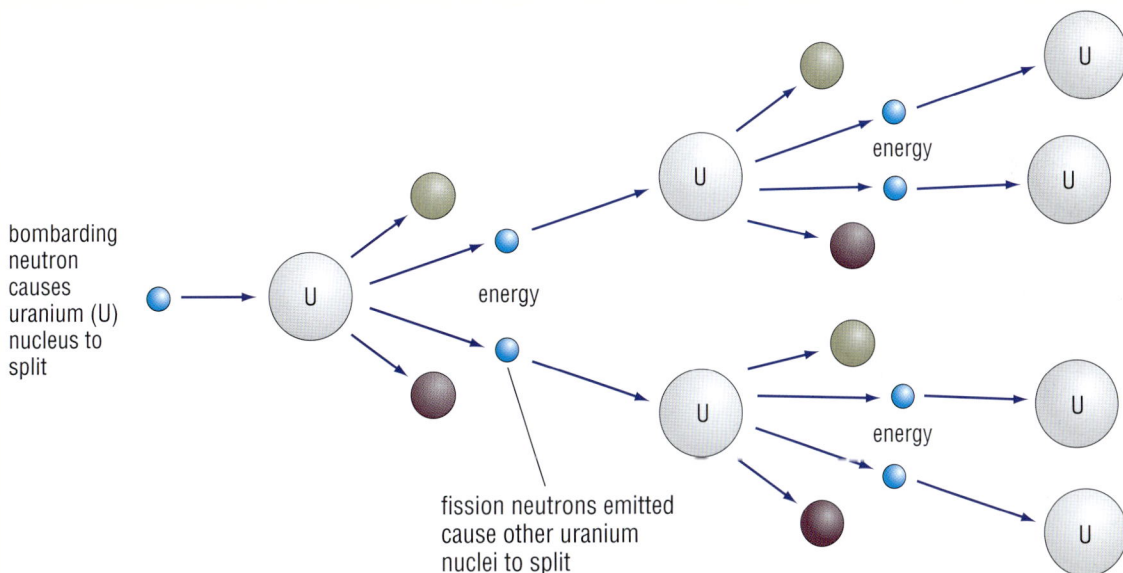

bombarding neutron causes uranium (U) nucleus to split

energy

fission neutrons emitted cause other uranium nuclei to split

Figure D   A chain reaction.

The nuclear reactor was developed to release energy in sufficient amounts to heat water to steam. This could then be used to spin a turbine in a generator to produce electricity. Today there are about 400 nuclear reactors operating in more than 30 countries.

   Nuclear reactors are built and operated to very strict safety rules to prevent them overheating and exploding. They do not produce sulphur dioxide and carbon dioxide like the power stations using fossil fuels, but the nuclear wastes that they produce have to be stored for thousands of years while they decay to harmless materials.

For discussion

Discuss this statement: 'The study of radioactive materials has been a great benefit to humankind'.

For discussion

Can a house be designed which would get all its energy from solar panels and solar cells? In the design there can be batteries to store electrical energy.

For discussion

Can a small town be designed to run on energy from solar panels and solar cells?

Renewable energy sources

These energy resources can be replaced in a short time and so are not used up.

Solar energy

We can use two forms of solar energy as sources of renewable energy.

Heat energy can be trapped by solar panels (see Figure 15.9) and used to heat buildings.

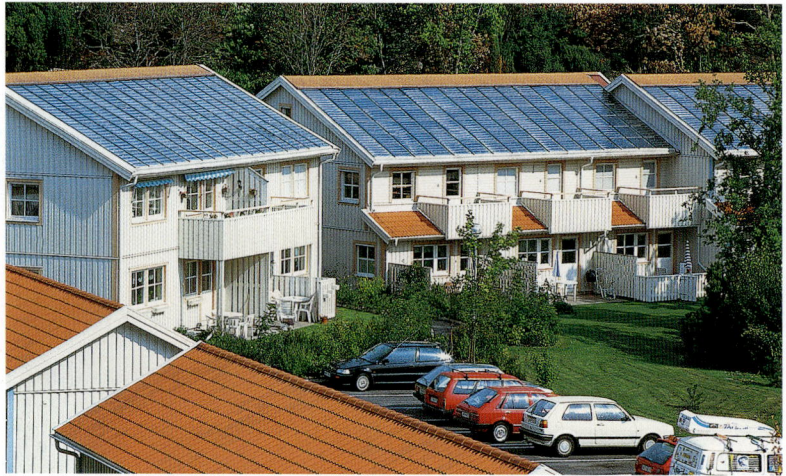

Figure 15.9   These roofs in Sweden are fitted with solar panels.

Light energy can be converted into electric energy by solar cells (see Figure 15.10).

Figure 15.10   Arrays of solar cells project from the International Space Station to collect some of the Sun's energy.

9 Draw an energy transfer diagram for a generator using the energy in a wave.

10 What expenses would have to be met to set up machines to generate electricity from waves and transport the electricity to the shore?

Waves as energy sources

Waves move up and down as they pass by. Machines are being developed to convert the up and down motion of the wave into a circular motion, which can turn the shaft of a turbine and generate electricity. The machines need to be moored in places where the waves are frequent but not too strong to destroy them. Figure 15.11 shows a machine being brought in shore after trials at sea.

Figure 15.11 This machine (here being towed back from sea trials) converts the energy in the up-and-down motion of passing waves into a turning motion required to make an electric generator work.

Rivers as energy sources

The kinetic energy in moving water in a river is a source of renewable energy. However, the flow of water for providing useful energy for people needs to be controlled. If a turbine was simply lowered into a river and left there, it would spin when the river flowed normally, it would hardly turn at all when the river had little water in it and it might be washed away when the river was in flood. This was known by the early engineers who set up water mills. They also set up a means of controlling the flow over the water wheel. The water was released steadily from a dam with a lodge or lake behind it.

Today dams are built to store huge amounts of water to turn the turbines in hydroelectric power stations (see Figure 15.12 overleaf).

For discussion

Large dams are built to provide the energy for large power stations (see page 98). When a large dam is built a large reservoir forms behind it, as Figure 15.12 shows. Often there are villages on the land that is to be flooded. Imagine you have to explain to the leader of a village that the people in the village will have to move. You have to make the movement of villagers out of the area as peaceful as possible. You can offer them incentives to move such as money and land but you must try and move them as cheaply as possible.

You may like to develop this activity into someone taking the role of the village leader and others taking the role of villagers while you and a few others take the roles of people representing the company building the dam and power station.

Figure 15.12 Water released from this dam is used to turn turbines connected to generators in a hydroelectric power station.

Wind as a source of energy

Just as there has to be a steady flow of water for harnessing energy from the current, so there has to be a steady flow of wind for a wind turbine to work efficiently. Obstacles such as buildings can cause turbulence in the air when the wind passes over them and if the turbine is sited near a building it may not get a steady flow of the wind to keep it turning smoothly.

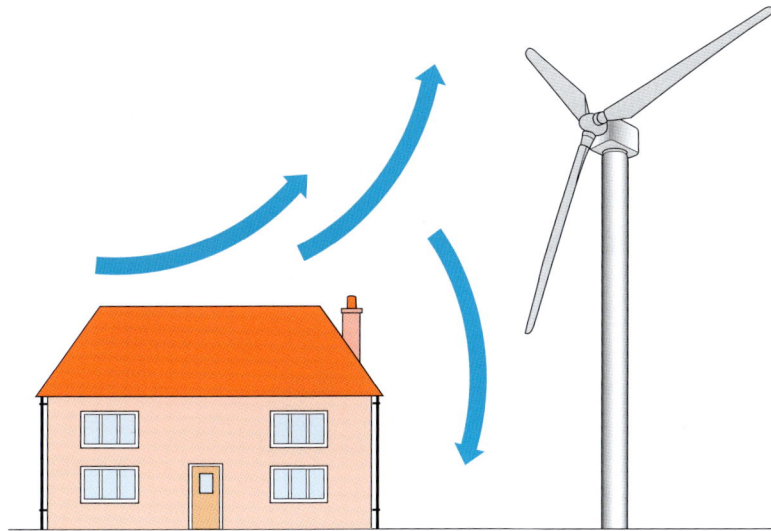

Figure 15.13 House near turbine.

When wind blows against a cliff face, turbulence develops at the top. If a turbine is sited at a clifftop it will not turn smoothly.

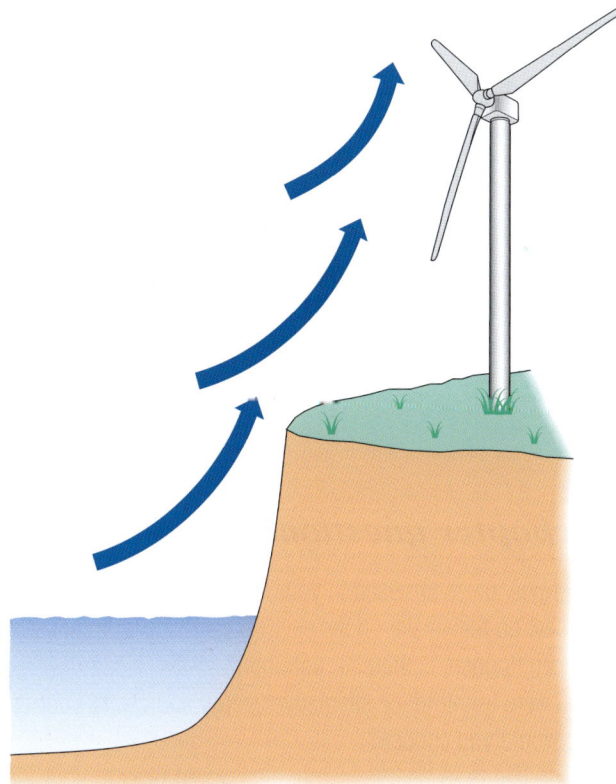

For discussion

If you have not tried the activity in the introduction (page 6) about a wind farm you may like to try it now.

Figure 15.14 Turbine on clifftop.

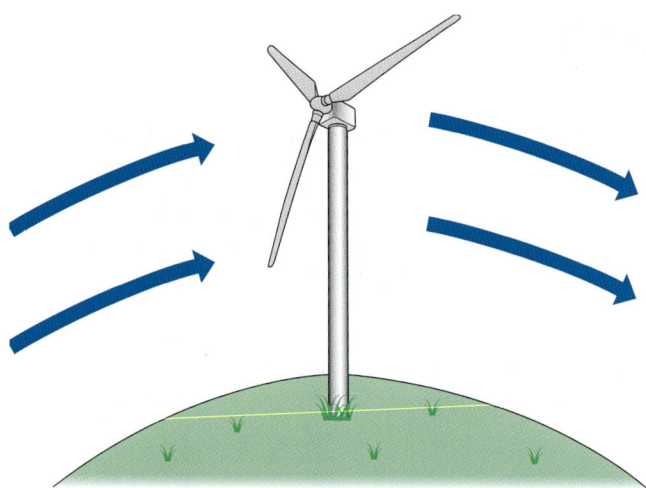

Figure 15.15 Turbine on hilltop.

A hill which raises steadily to a rounded hilltop then sinks slowly on the other side is a good site for a turbine. It should be set up at the top of the hill. The wind here will move smoothly over the hilltop like the air on an aircraft wing and keep the turbine turning steadily.

However, wind is not the only consideration in siting a wind farm. Turbines also make a noise. As the strength of the sound dies away with the increasing distance from the source of the sound, turbines need to be sited away from habitations.

◆ SUMMARY ◆

◆ Some of the Sun's heat energy drives the Earth's water cycle (*see page 161*).
◆ Some of the Sun's heat energy generates the winds and waves (*see page 161*).
◆ Energy pathways on Earth can be shown in a Sankey diagram (*see page 162*).
◆ Fossil fuels store energy from the Sun (*see page 163*).
◆ The world today depends on fossil fuels for energy (*see page 164*).
◆ There are problems with the extraction of coal, oil and gas from the ground (*see pages 164–165*).
◆ The price of fuel is related to the cost of extracting it (*see page 166*).
◆ Burning fossil fuels can cause damage to the environment (*see pages 166–167*).
◆ Radioactive elements release energy as they decay (*see page 167*)
◆ Energy resources which can be replaced are called renewable energy sources (*see pages 170–174*).

End of chapter questions

1   What energy sources are used in your region?
2   What impact do they make on the environment? For example, have valleys been flooded, is there air pollution?
3   **a)** If you were to introduce alternative energy sources to your area what would you choose?
    b) How do you think people in your area would react to your ideas?

Glossary

Some of the words in this glossary are pink. These are words which feature in the wordlists of the Checkpoint Science (Physics) Scheme of Work.

A

absorption   The process in which something is taken in by an object. For example, a sponge taking up water or the surface of a car tyre taking up light.

acceleration   The change in velocity of an object in a certain period of time.

air resistance   The backward push of the air on an object moving through it.

alternative energy source   An energy source which can be used as an alternative (instead of) to a source already in use. For example, wind is an alternative energy source to coal.

ammeter   An instrument for measuring the size of a current flowing through a circuit, in amperes.

amplitude   The maximum displacement of a vibrating object from its rest position.

area   The measure of a surface.

atom   A particle of an element which comprises a central nucleus that is surrounded by electrons.

B

battery   Two or more electrical cells joined in series in a circuit.

bimetallic strip   A strip of material made from two different metals which differ greatly in the way they expand and contract; used in thermostats.

C

cell   A device containing chemicals which react and produce a current of electricity in a closed conducting circuit.

centripetal force   The force that acts on an object moving in a circle, pulling it in towards the centre of the circle.

charge   There are two kinds of charge – positive and negative. A proton in an atom has a positive charge and an electron has a negative charge. Positive and negative charges can be built up on certain materials, called insulators, when they are rubbed.

chemical energy   The energy stored in the links between atoms of a substance.

circuit   An arrangement of wires and electrical devices around which a current of electricity can flow.

circuit breaker   An electrical device that stops the flow of electricity through a circuit if the current becomes too high to pass safely.

compass   A scientific instrument used to find magnetic north and south; it also shows other directions such as east and west.

conduction   The passage of heat energy from one part of a material to another, by vibrating particles passing kinetic energy on to neighbouring particles.

conductor (electrical)   A material that allows electricity to pass easily through it.

conductor (thermal)   A material that allows heat to pass easily through it by conduction.

conservation   a) of energy – the keeping of the same amount, b) of resources – the saving of resources for future use, c) of wildlife – maintaining natural habitats.

convection   The passage of heat energy through a liquid or a gas, by the particles in the substance changing position and carrying kinetic energy with them.

current   A movement of electrically charged particles.

D

density   The mass of unit volume of a substance.

diode   An electrical component which allows a current to pass in only one direction through it.

dispersion   The spreading out of light of different colours from a beam of sunlight.

domain   A tiny region inside a magnet which behaves like a microscopic magnet.

drag   A force which acts on a moving body in the opposite direction to which the body is moving, due to air or water resistance.

dynamo   A device for generating a current of electricity. There are two kinds: a direct current or d.c. dynamo which produces a current that flows in only one direction, and an alternating current or a.c. dynamo which produces a current that changes direction many times a second. The bicycle dynamo and power station generators are a.c. dynamos.

E

elastic limit   The maximum force that can be applied to an elastic material without the material becoming permanently deformed.

elastic material   A material that exerts a strain force when deformed, tending to restore it to its original shape.

electromagnet   A magnet which is made by coiling a wire around a piece of iron then passing a current through the wire.

electromagnetic waves   Waves with electrical and magnetic properties which transfer energy, such as light and radio waves.

electron   A tiny particle inside an atom which moves around the nucleus. It has a negative electric charge.

electrostatic   A condition in which large numbers of electrical charges stay in one place.

electrostatic charge   A charge of electricity which stays in place on the surface of a material; it may be positive (due to a lack of electrons) or negative (due to an excess of electrons).

energy   The ability of something to do work.

energy converter   A material or an object in which energy changes from one form to another as it passes through; also called an energy transducer.

evaporation   A process in which a liquid turns into a gas without boiling.

expansion   Enlargement of a material due to its increase in size (or space that the material occupies).

F

field   A region in which a non-contact force acts.

force   A push or a pull; it may be a contact force, for example an impact force, or a non-contact force, for example a magnetic force.

fossil fuel   A fuel such as coal, oil or gas which is formed from the fossilised remains of plants or animals.

frequency   The number of waves passing a point in a certain amount of time.

friction   A force that acts against the relative movement of two surfaces in contact.

fuel   A substance used to provide energy for heating, producing electricity or working machinery.

fuse   A wire with a low melting point which is placed in a circuit to prevent high currents flowing through it.

G

geothermal energy   Energy extracted from hot rocks beneath the surface of the Earth.

gradient   A sloping surface such as a ramp.

gravity   The force of attraction between any two masses in the Universe. The force is noticed when the two masses are very large, for example planets, or when one is very large and the other is very small by comparison, for example the Earth and you.

H

heat (thermal) energy   The energy transferred to or from a substance by heating, which increases or decreases the internal kinetic energy of the substance.

heat energy transfer   The movement of heat energy through a material.

hydraulic system   A machine made from pistons and pipes that contain a liquid. It transmits pressure and converts a small force into a large one.

hydroelectric power   Electricity produced from the energy of falling water.

I

image   The picture of an object which is produced when light is reflected from a mirror or is focused onto a screen by a lens.

impact force   The force exerted by one object on another when they collide.

induced charge   An electric charge which develops on the surface of a material due to the presence of an electrically charged object close by but not in contact.

induced magnetism   Temporary magnetism produced in a magnetic material when it is close to a magnet.

insulator (electrical)   A material that does not allow electricity to pass through it.

insulator (thermal)   A material that does not allow heat to pass easily through it by conduction.

internal energy   The energy that atoms of a substance possess, partly due to their motion.

K

kinetic energy   The energy possessed by a moving object.

L

lamp   An electrical device which converts some electrical energy into light energy.

LDR   Light dependent resistor.

LED   Light emitting diode.

lightning   A rapid movement of a large amount of electrical charge through air which produces heat and light.

longitudinal   A direction which runs along the length of an object or path, not across it.

luminous object   Any object that releases energy in the form of light.

M

magnet   An object which can attract magnetic materials and point in a north–south direction when it is free to do so.

magnetic field pattern   The arrangement of iron filings when allowed to settle freely on a card above a magnet.

magnetic material   A material that is attracted to a magnet and can be made into a magnet.

magnetic pole   One of two regions in a magnet where the magnetic force is very strong.

mass   The amount of matter in an object.

motion   A word used to explain a kind of movement. For example, the forward motion of a train or the flapping motion of a bird's wings.

N

non-luminous object   An object that does not release energy in the form of light but may reflect light from luminous objects.

non-magnetic material   A material that is not attracted to a magnet and cannot be made into a magnet.

non-renewable energy source   A source of energy, such as fossil fuels and radioactive materials, which cannot be replaced once it has been used.

nuclear energy   The energy stored in the nucleus of an atom.

nuclear fission   The process in which the nucleus of an atom breaks down into smaller nuclei and releases energy.

nuclear fusion   A process in which atomic nuclei join together to form larger nuclei of other elements.

nuclear reactor   A device in which nuclear fission is allowed to take place safely so that the energy released can be used to generate electricity.

O

opaque material   A material through which light cannot pass.

P

parallel circuit   A circuit in which some components are set side by side and the current flows in parallel paths through them.

particle   A very small piece of matter. The word can be used to refer to grains of sand, atoms in materials and tiny objects inside atoms such as electrons.

phenomenon   Something which can be observed that is due to the way matter behaves; for example when a piece of wood gets very hot it bursts into flame and gives out light.

pitch   A measure of the frequency of a sound wave.

potential difference   The difference in electric potential between two points of a circuit, which is the cause of the current flow. The current flows from a point of higher potential to a point of lower potential.

potential energy   Energy that is stored when work is done moving an object against a force. It is the energy something has due to its position, for example a person who has climbed to the top of a slide has a larger amount of potential energy than someone who has just slid to the bottom.

pressure   The term used to describe a force acting over an area of known size.

primary colours (of light)   Red, green and blue. They can be used to make all the other colours of light.

prism   A piece of transparent glass or plastic which has a triangular cross-section; used to disperse the coloured light in sunlight to form the visible spectrum.

R

radiation   A form of energy transfer by electromagnetic waves.

radioactive materials   Materials in which nuclear reactions take place and energy is released in the form of nuclear radiation.

reaction force   If object A exerts a force on object B, object B exerts an equal and opposite reaction force on object A.

real image   An image that can be focused onto a screen.

reflection   A process in which light rays striking a surface are turned away from the surface.

refraction   The bending of a light ray as it passes from one transparent substance to another.

renewable energy source   A source of energy such as sunlight, wind and biomass, which can be used again and again.

repulsion   The process in which two similar magnetic poles push each other apart.

resistance (electrical)   The property of a material which opposes the flow of a current through the material.

resistor   A device that offers a certain amount of resistance to a current passing through a circuit.

S

scale   A line on which units of measurement are marked. For example, the millimetre scale on a ruler.

scatter   To move away in all directions from a certain point.

series   An arrangement of electrical components in a line.

shadow   The dark area without light which forms behind an object when light is shone onto the front of the object.

sliding friction   The friction that exists between two objects when one is moving over the other.

solar   Energy coming from the Sun or objects which use its energy (e.g. solar panels).

solar cell   A device that converts the energy in sunlight into electrical energy.

Solar System   The Sun and the planets, moons, asteroids and comets which move around it.

sound energy   The energy transferred by a sound wave.

spectrum (electromagnetic)   The full range of wavelengths of electromagnetic waves from the shortest gamma rays to the longest radio waves.

spectrum (visible)   The bands of coloured light seen when a prism disperses sunlight. The colours are red, orange, yellow, green, blue, indigo and violet.

speed   A measure of the distance covered by a moving object in a certain time.

standard form   The standard system of recording large numbers, with only one figure in front of the decimal point, for example 3.84×10^8.

static friction   The friction that exists between two objects when there is no movement between them. It acts against an applied force, preventing movement.

stored energy   Energy which is not moving but when it is released some of it can be changed into moving energy.

strain energy   The energy stored in an elastic material when it is deformed.

strain force   The force exerted by an elastic material when it is deformed; it acts in the opposite direction to the applied force.

streamlined shape   A shape that allows an object to move easily through air or water.

switch   An electrical device which allows a current to flow but can also stop it flowing. When the switch is closed (on) a current flows. When the switch is open (off) the current cannot flow.

T

temperature   A measure of the hotness or coldness of a substance; it depends on the average kinetic energy of the particles.

terminal velocity   The velocity at which an object falls through air when the air resistance balances the weight of the object.

transformation   A process in which something makes a complete change from one form to another.

transformer   An electrical device that changes the voltage of the electricity.

translucent material   A material that allows some light to pass through it but scatters this light in all directions.

transparent material   A material that allows light to pass through it without the light being scattered.

U

ultraviolet   Electromagnetic waves with wavelength shorter than blue light.

unit   A standard for measurements, for example the kilogram.

upthrust   The upward force exerted on an object by the liquid or gas around it that it displaces. The force is equal to the weight of the displaced liquid or gas.

V

vacuum   A space in which there is no matter; it contains no atomic particles.

velocity   The speed of an object or a wave in a particular direction.

vibration   The rapid movement of an object to and fro about a rest position, as seen when a guitar string has been plucked.

virtual image   An image such as the one seen in a mirror which cannot be focused onto a screen.

voltage   The difference in electric potential between two points such as the terminals of a cell, measured in volts.

voltmeter   A device that measures the difference in electric potential between two parts of a circuit, in volts.

volume   The space occupied by a certain amount of matter.

W

weight   The gravitational force between an object on a planet such as the Earth and the Earth itself. The weight pulls the object towards the centre of the Earth.

work   The energy expended when a force moves an object through a distance.

X

X-rays   Electromagnetic waves with wavelength between ultraviolet light and gamma rays.

Index

Note: page numbers in *italics* refer to entries in the Glossary.

INDEX